Designing Modern Britain

Designing Modern Britain

Cheryl Buckley

REAKTION BOOKS

Published by Reaktion Books Ltd
33 Great Sutton Street
London EC1V 0DX, UK

www.reaktionbooks.co.uk

First published 2007

The publishers gratefully acknowledge support for the publication of this book by
The Paul Mellon Centre for Studies in British Art.

Printed and bound in China

British Library Cataloguing in Publication Data
Buckley, Cheryl, 1956–
    Designing modern Britain
    1. Design – Great Britain – History – 20th century
    I. Title
    745.4'0941'0904

ISBN–13: 978 1 86189 322 2
ISBN–10: 1 86189 322 1

Contents

FLY YOURSELF . . . FLY
YOUR FREIGHT to fifty-one
countries on all six continents by
B.O.A.C. Also within Britain and
from London to the principal
cities of Europe by B.E.A.

B·O·A·C / BEA

BRITISH OVERSEAS AIRWAYS CORPORATION. AIRWAYS TERMINAL,
BUCKINGHAM PALACE ROAD, LONDON, S.W.I. TELEPHONE: VICTORIA 2323

BRITISH EUROPEAN AIRWAYS, DORLAND HOUSE,
REGENT ST., LONDON, S.W.I. TELEPHONE: GERRARD 9833

BEA

Introduction

Change and continuity characterised design in twentieth-century Britain. Framed by de-industrialization, political and economic realignment and nostalgia for selected periods of 'English' history, design in Britain also articulated 'the maelstrom of modern life'.[1] Ascendant for a significant part of the century, modernist discourses eschewed historical styles and decoration in favour of a new visual language inspired by geometry and abstraction, mass-production and the aesthetics of the machine, but British responses to twentieth-century modernity and modernism proved distinctive and remarkably varied.[2] Defining design not just as 'things' but also as a matrix of interdependent practices, *Designing Modern Britain* considers the ways in which it represented and constructed modernity at crucial moments from the end of the nineteenth century to the beginning of the twenty-first. Integral to a capitalist economy that was in transition, design provided an effective tool in Britain's self-fashioning by referencing the past, present and future.

Designing Modern Britain shows how design responded to regional and local exigencies, as well as to international initiatives. Articulated as singular and homogenous by various proponents, modernism was transmuted across Britain as it interacted with local traditions, responded to historical precedents, reworked decorative and figurative idioms, adapted to manufacturing conditions and priorities, and took account of consumer needs and desires. To the critic Roger Fry, writing in 1913, 'Englishness' in design encapsulated simplicity, ruggedness and naivety, qualities he also found in modernist art and non-Western cultures, whereas Nikolaus Pevsner cited 'un-Englishness' as an essential feature, thus highlighting the important contribution that émigrés and immigrants made to the complexity of design in twentieth-century Britain.[3]

Advertisement for BOAC/BEA (British Overseas Airways Corporation and British European Airways), Festival of Britain, South Bank Exhibition, 1951.

The ISOKON LONG CHAIR is light and well balanced. It can be moved and lifted easily (even by a child). It offers, with its surprising mobility, new ways of arrangements, even in a limited space.

You can sit and turn sideways by a fire instead of burning your feet while your body freezes. Two can be placed side by side but facing each other—an ideal arrangement for a quiet talk, or for tea. They can be moved to the window to get the sunshine.

You cannot do any of these things with heavy, old-fashioned chairs.

TYPOGRAPHY BY MOHOLY-NAGY
PRINTED BY LUND HUMPHRIES

Isokon Furniture Co. sales leaflet designed by László Moholy-Nagy, c. 1936.

Rather than acting as a comprehensive history of British design or providing a survey of its numerous categories (engineering, transport, graphics, furniture, architecture, industrial design, product design, fashion, ceramics or textiles), *Designing Modern Britain* explores the connected themes of modernity and identity through selected case studies. To be modern, progressive and forward-looking were essential as Britain negotiated its world position, and as a vital tool in economic recovery, design responded to global competition from Europe, the USA and more recently Asia. In the 1930s the Design and Industries Association aimed to 'induce British manufacturers to improve the design of their goods' so as to meet the demand from the general public for better designs at inexpensive prices, which at the time was 'being largely met by imports from abroad',[4] whereas in 1947 the President of the Board of Trade, Stafford Cripps, argued: 'design is of crucial importance to British industry to-day, whether we think of what is due from our own producers to our own people, or how best to meet the very live competition which is ahead of us in foreign markets'.[5]

Manufacturers, designers and retailers all played important parts in promoting British products, as did writers, journalists, educationalists and theorists. Articles and features appeared in colour supplements, art, design and architectural periodicals and even children's literature, while critics and campaigners published manuals on taste and design, organized exhibitions and formed committees to improve, educate and cajole consumers, designers, retailers and manufacturers.

The parameters of 'modern' design have shifted over the course of the past century. As Britain's early modernists veered towards the eternal in their understanding of modernism (Fry looked for the immutable and the abstract in early English ceramics, whereas the designer and architect Wells Coates searched for 'the essential intention', for example), others focused on the contingent. The use of natural and/or indigenous materials, such as wood, brick and ash glazes, was one way to signal this, but important too were visual iconographies that registered particular identities, local, regional and national, as well as those relating to gender, class, race and generation. Modernism in Britain appeared rigid, authoritarian and patrician at various points in the twentieth century, but its overly didactic stance can be partly explained by its marginal institutional presence across Europe in the 1920s, its precarious standing in Britain in the 1930s, the paucity of actual design commissions, and its predominantly avant-garde stance disseminated by only a handful of sympathetic journals. A dogmatic tone was reinforced with the arrival of some émigré architects and designers, particularly those from the German Bauhaus, who, unsurprisingly, were defensive of ideas and practices recently denigrated and curtailed in 1930s Germany. Of course, some concession to the consumer was necessary, as the Isokon Furniture Company's sales leaflet noted: the design of the innovative 'Long Chair' required an understanding of the gendered consumer, as well as a 'rational' analysis of form and use. The modernist analysis of the conditions of modern life proved inspirational to many younger writers, designers, architects and educators in Britain, but from the 1960s the desire to be singularly 'modern' underwent revision with an increased acknowledgement of diversity and difference, as well as a refusal of core modernist premises. Some areas of design were particularly effective vehicles for this, and proved both constitutive of and reactive to postmodernism's polyvalence. Fashion and advertising – responsive to pop cultures and cultural eclecticism – engaged with

Advertisement for Marconi Wireless Telegraph Co. Ltd, Festival of Britain, South Bank Exhibition, 1951.

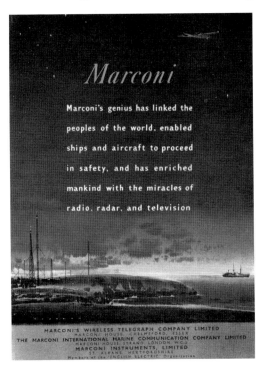

complexity and contradiction via retro styles, sub-cultural iconographies, second-hand and DIY aesthetics.

Designing Modern Britain questions the extent to which there was a distinct break from modernism to postmodernism in the last quarter of the twentieth century, but it also proposes that if being modern was critical, then so too was remaining 'English'. To Pevsner, 'Englishness' comprised contradictory qualities: reasonableness and rationalism on the one hand, and imagination and fantasy on the other,[6] but for the contributors to the DIA *Quarterly Journal* in 1929, England had begun to imply 'old England' with 'everlasting reproduction of century-old designs' and 'the cult of the antique'.[7] Some thirty years later, a feature on tourism in the *Sunday Times Colour Section* not only used Britain and England interchangeably, but had a highly selective 'take' on Britain's constituents: 'There is Wales; not the Wales of the great steelworks in the South but the Eisteddfod Wales, with women in comical conical hats. There is Scotland, which means kilts blowing on the parade ground of Edinburgh Castle.'[8] But mainly Britain meant England, delineated by a handful of historic towns and cities and forming a register of self-representation vital for economic reasons, but honed down to apparent 'essences': London, Stratford-upon-Avon, Oxford, perhaps the Lake District, but rarely Northumberland. Predominant representations of 'England' were synonymous with past glories, and in design terms particularly important was 'globally adventuring Tudor England'.[9] But economic priorities stimulated a desire to be forward-looking, particularly as Britain's political and economic power diminished through the twentieth century. As the Minister of Health, Arthur Greenwood, put it in 1929 – referring to England, but meaning Britain – what was required was 'an England of the twentieth century which shall be worthy of the best achievements of the past'.[10] In a context of economic under-performance, mid-century insecurities about the nation's world standing and emerging modernism, German ships were noted for their 'cleanness of line and consistency of decoration', whereas in British ships

> the carpets and upholstery looked as though the London warehouses had been ransacked for out-of-date stock that could be bought at reduced terms . . . and there was not the slightest indication in any detail, either in lighting, furniture or textiles, that she had been built to-day.[11]

Economics provided an important context for debates about Britain, modernity and design. This was apparent both before and after the Second World War. Except for a few deliberate journeys (J. B. Priestley in the 1930s, Bea Campbell in the 1980s and Pevsner beginning in the 1950s), the

Designing Modern Britain

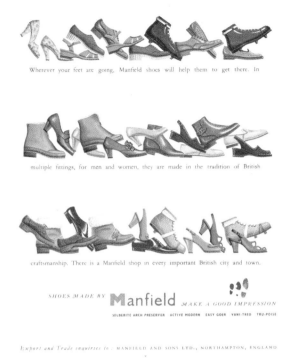

Wherever your feet are going, Manfield shoes will help them to get there. In

multiple fittings, for men and women, they are made in the tradition of British

craftsmanship. There is a Manfield shop in every important British city and town.

SHOES MADE BY Manfield *MAKE A GOOD IMPRESSION*

SELBERITE ARCH PRESERVER   ACTIVE MODERN   EASY GOER   VANI-TRED   TRU-POISE

Export and Trade enquiries to : MANFIELD AND SONS LTD., NORTHAMPTON, ENGLAND

Advertisement for
Manfield & Sons Ltd,
Festival of Britain, South
Bank Exhibition, 1951.

'English' regions, Northern Ireland and Wales, remained largely 'other' to these essential representations.[12]

The Festival of Britain in 1951 underlined the complexities and contradictions of design to date, but it was also a harbinger of things to come. Its organizers believed that it offered 'a new sort of narrative about Britain: an Exhibition designed to tell a story mainly through the medium, not of words, but of tangible things'.[13] This narrative, unusually focused on Britain rather than England, was mapped out in an official guide and in the design and layout of the exhibition pavilions. Overall, the Festival was an odd mix of the progressive, the paternalistic and the anachronistic, and this was particularly evident in the 'Lion and the Unicorn' pavilion. In an accompanying commentary, the writer Laurie Lee identified two apparently dominant national characteristics – realism and strength on the one hand, and fantasy, independence and imagination on the other. Developing this, Lee found clues to the British character in the long tradition of craftsmanship, 'in their old furniture . . . sporting guns and fishing tackle and tailoring', and in the British view of nature, which expressed itself in landscape and in the applied arts.[14] Thus far the story was the 'official one' – reiterating and reinforcing the notion that national identity resides in tradition and history – but the commercially sponsored guide, comprising 25 per cent advertising, suggested alternative stories by emphasizing new technologies and personal and domestic consumption. Equally, technology-led companies such as Napier turbo-chargers, Marconi television, Costain Construction, EMI electronics and Siemens telecommunications looked beyond Britain; as one advertiser put it, 'Marconi's genius has linked the world'. Implicit was the need for British companies to square up to foreign competition, but of equal significance was the domestic consumer market. From the evidence of the guide, personal and domestic goods were targeted at women in particular; shops and shopping (Barkers of Kensington, Liberty's of Regent Street), fashion and accessories (Manfield Shoes, Kayser Bond Lingerie, Daks Fashion), electric appliances (Hoover, Creda, English Electric),

Faith has been kept

IN THE YEAR 1860, an enterprising youth of sixteen began to sell horseradish to his neighbours. He had gathered it from his mother's garden and had conceived the then revolutionary idea of bottling it in clear glass so that all could see how good it was.

His name was H. J. Heinz.

Nine years later, when he formed the little food firm that has since grown into the world-famous organisation bearing his name, this was its foundation stone :

"To make a thing that the world needs, to make it the best that it can be made, to get it into the hands of the largest number of people at a fair price, and to charge a reasonable fee for the service."

That is what he believed. That is what he practised: all his life. For the past 55 years, the British House of Heinz has kept faith with his ideals. With the result that probably the most widely known and trusted foods in this country are the "57 varieties"...requiring the output of two model factories—working to capacity —to meet the home and overseas demand for them.

H. J. HEINZ COMPANY LIMITED,
LONDON, N.W.10.

THE BRITISH HOUSE OF **HEINZ** 57

textiles (Sanderson), cooking utensils (Prestige), dental care (Maclean and Addis), and food, drink and cigarettes (Heinz, Ovaltine, Cow & Gate, and Capstan). Largely figurative and narrative in style and tone, these images evoked a variety of 'Englands': cosy and rural, but also industrial and increasingly Americanized. Some emphasized streamlined modernity and efficiency, while others stressed the global nature of the company (Rootes Motors Ltd, Thomas W. Ward, Iron and Steel, Ferguson Tractors and BOAC/BEA). A number of advertisements pointed to the penetration of the domestic market by foreign goods. Paradoxically, enterprise, innovation and historical continuity were highlighted by Heinz and Hoover; one, a mid-nineteenth-century food-processing and canning manufacturer, the other an early twentieth-century domestic appliance manufacturer – but both US companies. 'The British House of HEINZ', to paraphrase the advertising copy, was based on the enterprise of its nineteenth-century originator, a 16-year-old youth who had brought convenience, mass availability and quality to the Victorian diet. A hundred years later Hoover enabled 'progress in the home' via the use of the Hoover vacuum cleaner and electric washing machine. Unquestionably patriarchal, such advertising depicted 'wife' and 'mum' working in a bright, modern kitchen undertaking household chores made easy with the assistance

Advertisement for H. J. Heinz Co. Ltd, Festival of Britain, South Bank Exhibition, 1951.

Advertisement for Hoover Ltd, Festival of Britain, South Bank Exhibition, 1951.

Designing Modern Britain

Cussons

IMPERIAL 🎖 LEATHER

The Quality Toilet Soap everyone wants!

Advertisement for
Cussons Sons & Co. Ltd,
Festival of Britain, South
Bank Exhibition, 1951.

of new technology that was not only 'best in design, best in materials, best in quality of workmanship', but responsive to consumer needs, with 'a model suitable for every size and type of home'.[15] The ideal consumer for such products was affluent middle-class, female and white. Britain as represented in this Festival guide was predominantly 'white'; yet migration had characterized modernity as people moved from Britain's former colonies overseas often in search of work. 'Blackness' appeared only as a guarantee of 'whiteness', as evident in the Cussons Imperial Leather soap advertisement in the guide.

Non-European or non-Western cultures were hugely significant influences on design. In part promoted by modernist theories, this was perhaps more implicit in the first half of the century, whereas in the second half it became explicit, since immigration led to the development and synthesis of different cultural agendas. Intersecting with those from other Western countries, particularly from Europe but increasingly the USA, and combined with notions of 'Englishness' and/or 'Britishness', these undercut the apparent hegemony of modernist theories and practices. *Designing Modern Britain* offers an insight into this complex matrix, and by acknowledging the permeability of design cultures, it highlights the contingent relationships between design, modernity and identity so as to examine a persistent aim: to go modern, but remain British.

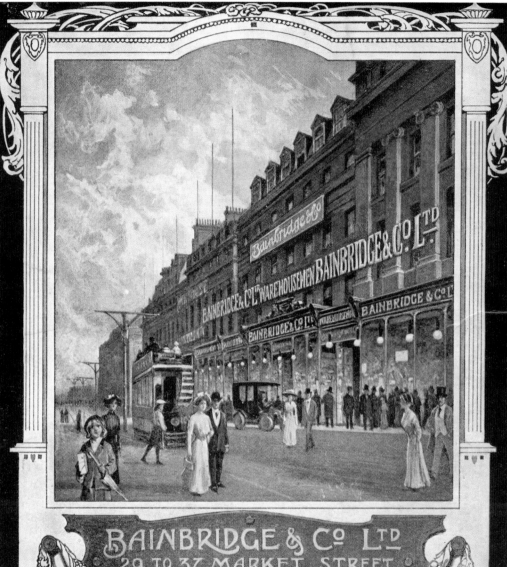

BAINBRIDGE & CO LTD
29 TO 37 MARKET STREET
& 26 & 28 BIGG MARKET.

1

Modernity and Tradition: Late Victorian and Edwardian Design

Between 1890 and 1914 the economic and social foundations of British life were subject to significant changes, but alongside continuity and stability. In terms of design, manufacturers produced goods in response to foreign competitors; large industrial combines began to adopt Taylorist and Fordist methods to improve efficiency, thus delivering different categories of design 'goods' to a larger and more broadly representative population (the census population figures were 41.5 million in 1901 and 45.25 million in 1911).[1] Some aspects of design helped to sustain and legitimize British imperial power, but others undermined it profoundly. The 'New Woman' clad in a mannish jacket, shortened skirt and straw boater was a potent symbol of women's demands for political representation and a material expression of the economic changes that were propelling some women into the public world of work, away from the private sphere of home. Conversely, the socially reformist Arts and Crafts Movement had lost some of its radical edge by the early twentieth century, but the influence of its writers and practitioners was diverse and far reaching; it affected art and design education, government economic policies (particularly via the Board of Trade), the attitudes and practices of manufacturers, the marketing and advertising strategies of retailers, and consumers. Consolidated before the First World War, aspects of Arts and Crafts design evoked stability, longevity and tradition; increasingly eclectic in its sources, it began to merge into the mainstream, drawing on styles from the cultural high points of British history: Tudor, Elizabethan Georgian and Regency. As it became increasingly important to distinguish 'British' design culture from European approaches, 'Englishness', 'Scottishness' and 'Irishness' were marshalled to differentiate 'British' from 'European' design, which was presented as alien, exotic and in some ways 'other'. However, the networks within which design was shaped

Publicity brochure for Bainbridge department store, Market Street, Newcastle upon Tyne, c. 1911.

in Scotland – in Glasgow and Edinburgh – were connected as much to Vienna, Darmstadt, Chicago and Turin as they were to London, and these were as concerned with modernity and internationalism as with tradition and national identity.

Design in late Victorian and Edwardian Britain was diverse and contradictory, as indicated, for example, by rebuilding in London. This resulted in grandiose 'imperial' vistas, including Aston Webb's Admiralty Arch of 1909, which deployed formal classical styles to symbolize 'the heart of Empire'.[2] Reinforcing this evocation of imperial power were the public displays found in department stores and national and international exhibitions – those of 1901 (Glasgow International), 1908 (Franco-British) and 1911 (Festival of Empire) – that fed 'spectatorial lust'.[3] They highlighted Britain's economic authority and pointed to the importance of the consumer within the culture of high capitalism.[4] This was in marked contrast to the suffrage displays and demonstrations that took place between 1907 and the outbreak of war in 1914, which used the design cultures of posters, banners, needlework, ceramics and jewellery to demand the vote rather than to stimulate consumption. In such a context of contradiction and uncertainty, the countryside as a symbol of national identity (more 'English' than 'British') came closer to the 'town' via the Garden City movement, but at the same time it became more isolated for a new generation of elite craftsmen and women who sought the 'apparent' simplicity of the English countryside, yet turned to the East for inspiration.

As several commentators have observed, England was but one part of the larger multinational state, the United Kingdom, which was characterized by differences in culture, language and religion (admittedly, England accounted for roughly three-quarters of its total population).[5] National identity was a particular concern, and the issue was highlighted by immigration. Immigration by Central European Jews prompted the Aliens Act of 1905; the question of Irish nationalism came to a head with the Home Rule Bill proposed in 1912; and the Indian Home Rule Society was founded in London in 1905 to rid India of the British. In such a context, cultural representations of empire played a crucial part in defining national identities.[6] Gendered and class identities mutated, partly due to the various reform acts beginning in the 1860s, which culminated in suffrage campaigns in the early 1900s. Domestic goods, fashionable dress and personal products were arenas in which a newly enfranchised populace (by 1910 this numbered 6 million) could exercise some degree of consumer choice, aided and abetted by advertising in the women's periodicals and the popular press; the *Daily Mirror* was re-launched in 1904 as a cheap (halfpenny) illustrated daily, and *Home Chat*, one of the new, cheaper women's magazines, was launched in 1895.

Designing Modern Britain

Historians of this period identify a handful of concerns that are also significant for design historians. The role of the state is one of these: housing and town planning, personal health and hygiene, and education all attracted state involvement.[7] At the same time, the individual was subject to contradictions: regulated and moulded, but at the same time informed of diverse habits and values.[8] Of concern to economic historians was the relative strength or weakness of the British economy. Britain's role as the lynchpin of the international economy was eroded from the end of the nineteenth century, as the economies of other countries, particularly the USA, emerged as highly efficient and strong.[9] By 1900 there were calls for a 'Fair Trade' rather than 'Free Trade' policy in response to the tariffs and import duties levied by Germany and the USA to protect their home markets. Britain had to compete not only abroad, but also at home, as imported manufacturing goods (tariff-free) were consumed by the growing domestic market. It was in the manufacture of these products – the result of the second industrial revolution, utilizing new technologies such as electricity – that the USA and Germany enjoyed increasing pre-eminence between 1890 and 1914. Britain, meanwhile, predominated in the production of staple goods – coal, textiles, iron and steel; in effect, the products of the first industrial revolution – whereas it was more sluggish in developing newer industries such as road vehicles, scientific instruments, advanced chemical products and electrical engineering. As Britain's staple industries declined, new sectors emerged that were linked to the service sector; these were consumption-led and based on mass-production techniques.[10] These newer industries were stimulated by the First World War, which reinforced scientific and technological progress.[11]

Perceptions of empire with regard to economic well-being were complex. Some have noted that the Edwardians had a negative perception of economic performance, which was exemplified by the complexities of maintaining the empire. But at a psychological level, the empire reinforced the idea of British superiority – its institutions and its people. It shaped people's perceptions of themselves in relation to the world as they travelled to the colonies in preference to Europe.[12] It stimulated knowledge and learning about animals, botany, geology, meteorology and anthropology. The 'imperial mentality' was summed up by a respect for hierarchy, order, militarism and a masculine view of the world, in marked contrast to domestic politics, which ran 'in quite the opposite direction, towards egalitarianism, "progressivism", consumerism, popular democracy, feminism and women's rights'.[13]

In some respects Britons appeared better off, because of the falls in the death, birth and infant mortality rates, but the wealthiest 1 per cent of the population became ever wealthier. 'Better off' encompassed not just material wealth but also social improvements, since the professional classes had half the number of children as their parents.[14] This was not the only aspect of change that benefited women. Although little of women's part-time work was recorded in the census returns, and the categories of work were too rigidly applied, in the 1901 census 77 per cent of women in the age group 15–34 years worked, some in 'respectable' employment as teachers, clerks, typists, saleswomen or nurses, but mainly in unskilled jobs in domestic service and factories. Working women were increasingly unionized, either in such bodies as The Women's Trade Union League (WTUL), founded in 1874, or in the large general unions that amalgamated from smaller craft trade unions in the first years of the twentieth century. Additionally, some of these women became radicalized in the context of the suffrage campaigns, which reached a peak between 1903 and 1914. Importantly, women's social roles were scrutinized and a variety of campaigns were organized to try to bring about change.[15]

Home Help: Domestic Technologies and Consumption

The private world of home and family was a crucial space for design in late Victorian and Edwardian Britain. In 1899 Thorstein Veblen drew attention to the home and women's role within it as a site for consumption and social emulation. He cited the importance of consumption in the processes of differentiation, since rival class identities were negotiated through the ownership and display of certain categories of goods and specific forms of leisure.[16] The dual processes of emulation and conspicuous consumption go some way to account for the status and meanings of a range of domestic designs, but equally important were the changes to middle-class women's lives, which affected both the consumption of design goods and their production. The private space of home and the personal domain of the body were sites where class, gender and national identities were articulated and contested via design things: luxury goods that facilitated leisure and emulation, such as clothes, furnishings, glass, ceramics and textiles, alongside practical equipment for everyday work in the home.

Contrary to popular myth, few middle-class women were leisured in the home. With a moment for themselves, they were expected to engage in 'female' pursuits such as sewing, popularly known as 'work'.[17] Even for those with one daily servant – the typical number for a middle-class family – the home was 'high maintenance', apparent from the plethora of advice to be found in the host of new women's magazines emerging at this time. Finding and keeping a servant proved increasingly difficult as young girls moved from household to household, or took jobs in factories, shops and offices, doing work that offered a more appealing method of earning a living. Equally, for

Designing Modern Britain

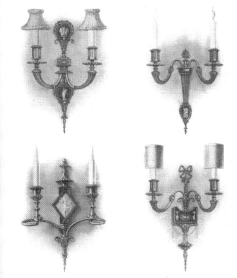

EXAMPLES OF WALL SCONCES ADAPTING 'ADAM' ORNAMENT
TO MODERN ELECTRIC LIGHTING REQUIREMENTS.

IN FINE CAST & HAND CHASED BRASS OF FRENCH GILT & OTHER
FINISHES & WITH WEDGWOOD CAMEOS OR ALL METAL ORNAMEN-
TATION :: ALL THE ABOVE & OTHER DESIGNS FROM OUR OWN M-
ODELS TO BE SEEN AT 112 VICTORIA S⊺ WESTMINSTER. S.W. (Art Fittings D⁹)
OR OBTAINABLE FROM THE MAKERS. NORMAN & ERNEST SPITTLE
TELEPHONE {724. WESTMINSTER : NEWHALL HILL · BIRMINGHAM.
         {2969. BIRMINGHAM :

Advertisement for
electric wall sconces,
1906. Modernity and
tradition were synthesized
as 'Adam' ornaments were
adapted to modern electric
lighting requirements.

those lucky enough to have a servant, it was increasingly common to supervise them closely, if not work alongside them.[18] These middle-class women were targeted by manufacturers and advertisers of the numerous products that claimed to help with housework, and although some clearly did, others merely created other forms of labour or contributed to 'housework inflation' – since there was increased expectation that houses should be cleaner, more comfortable and more aesthetically appealing. The twentieth-century home was similar in some respects to that of the nineteenth century, but some rooms gained in significance, such as the kitchen, thus contributing to new notions of 'home'.[19] Additionally, some types of domestic products were more popular than others, such as the sewing machine, the washing machine, the stove and the bath.[20] Contrary to the conventional image of the ornamental middle-class woman whose main acquisition was a piano, the sewing machine was an infinitely more useful and common possession, used for home dressmaking.[21]

It has been assumed that, as servant numbers dropped, women were increasingly assisted by new technologies (gas and electricity in particular), thus saving their labour in the home. In 1900 housework in a typical middle-class home included cooking, cleaning, heating and laundry – tasks that were time-consuming and exhausting, but which also required considerable planning, organization and management. Switching on an electric heater offered the potential for cleaner and swifter housework, whereas 'maintaining and cleaning a coal fireplace claimed a considerable amount of time'.[22] But products for cleaning, food storage, preparation and cooking equipment (including water heaters, floor sweepers, washing machines, cookers, irons and sewing machines) were slow to penetrate the middle-class household. For example, it was not until the mid-1930s that the electric iron was used in 50 per cent of British homes.[23] Indeed, in the cases of electric vacuum cleaners, washing machines, water heaters and refrigerators, it was not until the end of the 1950s that 50 per cent ownership began to be attained in Britain, whereas this occurred before the Second World War in the USA.[24] The slow take-up of electrical appliances that were 'time-saving' (equipment for cleaning, cooking, washing) was contrasted with electrical goods that were 'time-using' (radios). This was because there was a preference for goods that benefited both men and women in the home, such as radios and electric lighting.[25]

Whereas the ideal nineteenth-century middle-class home was meant to be a haven – attending to the aesthetic and moral needs of the family – the home in the early twentieth century was more concerned with health and cleanliness, and there was less clutter.[26] The health of the nation became a particularly contentious issue, since its deterioration – particularly apparent during recruitment for the Boer War – and high infant mortality rates contributed to anxiety over the physical and intellectual character of the country in the first decade of the twentieth century.[27] In such a context domestic hygiene became paramount, and efforts were concentrated in kitchens and bathrooms. Caught up in the rhetoric of cleanliness was the theme of rationalization: if only the kitchen could be like a factory, ordered and logical, planned and structured. Influenced by the writings of the American Frederick Taylor on scientific management, this concern for rationality emerged at a time when women's roles were being questioned within the context of the suffrage campaigns. Promotion of the home as a place that required technical planning and organization served as a means of enhancing the standing of the housewife at a point when gender roles were under scrutiny. But as the feminist Charlotte Perkins Gilman noted, women's equality would be attained as much by domestic design and reorganization as by the ballot box.[28] Kitchens in particular attracted a significant degree of attention, being simplified and streamlined. Taylorist time and motion studies – developed initially for mass-production systems in factories – were applied to work within the kitchen, as surfaces and equipment were arranged 'scientifically and ergonomically'. This marked an important change, because the role of middle-class women in the home became less that of an aesthetic arbiter and more of a skilled professional.[29] Women's new domestic skills, learned rather than intuitively acquired, were to bring efficiency to the kitchen, thus reducing the hours spent on housework. Time saved could then be used to raise standards and to improve the quality of the home.[30] This shaped not only domestic design, but also the emerging ideologies of domestic hygiene that led to the marketing of individual products – washing machine, iron, vacuum cleaner, cooker and water heater – to individual consumers. These new domestic appliances produced a particular type of domestic work in the home: isolated and based on the use of single appliances instead of the collectivist organization of domestic work envisaged by feminists such as Gilman. Undertaken by a lone housewife, these products reproduced a distinctive form of housework, but at the same time they secured a substantial middle-class market for the new goods.

In contrast, the vast majority of working-class women were unable to afford any of these goods. The patterns of their lives were governed by the avoidance of poverty, even though real incomes rose among the working class by 90 per cent between 1850 and 1914.[31] Working-class poverty was largely caused by insufficient wages. This was evident in Seebohm Rowntree's study of York undertaken in 1899 and published as *Poverty: A Study of Town Life* in 1901. In this, he differentiated between primary and secondary poverty – the former was due to insufficient earnings to support the minimum necessities of life (Rowntree termed this 'physical efficiency'), whereas the latter meant that, although earnings ensured physical efficiency, there was little for anything else, while factors such as ill health, family size, unemployment, alcohol problems or idleness might cause workers to move back and forth from primary to secondary poverty throughout their lives.[32] In such a context, what, then, was the capacity of the working-class family at the turn of the century to consume new products within the home?

For his survey, Rowntree visited the houses of different categories of working-class people, recording the conditions, facilities and belongings. In the houses of those categorized as 'well-to-do artisans' (representing 12 per cent of working-class families in York and earning more than 26s. a week), there was a sitting room (or parlour) with a piano, an overmantel, imitation marble fireplace and brightly tiled hearths, but the hub of the house was the kitchen, in which might be found a horsehair sofa, an armchair, china ornaments and polished tins. In the scullery were a sink, a tap and the copper. In Rowntree's case studies of families in this category, there were no appliances beyond the copper. In the houses of those with moderate, but regular wages (62 per cent of working-class families in York and earning under 26s.), pianos were less common, and often the parlour (if there was one) was used to store a bicycle or pram. Typically, the kitchen and parlour were combined, and in this room would be a table, two or three chairs, a wooden easy chair and perhaps a couch. A small scullery might contain a sink and copper, but there was often no tap. Atypically, one family had a wringing machine and a soldier's widow had a sewing machine.[33] A further 26 per cent of people lived in conditions that were poorer than those already described, in slums, back-to-back houses and lodgings; in these it was difficult to keep goods clean, although Rowntree observed that some did manage this. For example, a labourer earning 19s. a week and his wife (described as a good manager who took in washing and an occasional lodger) had barely any furniture, but their children looked bright and intelligent, and the food was nicely cooked. But typically furniture consisted of boxes, perhaps a couple of chairs, old and often dirty flock bedding – in interiors lacking ventilation. Children in these families were pale and half-clothed.[34]

Apparently conditions in York were typical, not exceptional. In *Round About a Pound a Week*, written in 1913, Maud Pember Reeves focused on the London borough of Lambeth, asking how working-class people lived on wages between 18s. and 30s. a week. What were their diets? What type of furniture did they have? How were they clothed? And how did they go about their daily lives?[35] Following Rowntree, she studied the housing and living conditions and noted furniture and property. Except for one woman who owned a treadle sewing machine, acquired before her marriage, there was little evidence of new technology in the home. Coppers were used for hot water, although they were often shared, and cooking was done primarily in coal ovens, with one or two examples of gas stoves in evidence. Significantly, working-class incomes meant that the use of gas was strictly monitored, so these stoves were used only for cooking Sunday lunch.

Rowntree's and Reeves's studies of the working class in York and Lambeth showed that the consumption of new technologies within the home at the end of the nineteenth century and the first decade of the twentieth was primarily a middle-class affair, but education was used as a tool to prepare working-class women for appropriate consumption. They were taught to invest in simple domestic technologies, such as gas cookers, a range of cutlery, pots and bowls, and simple washing and mangling machines from the 1880s.[36] Of those working-class women who had the income to 'invest', typically they chose labour-saving devices to raise the standards of cleanliness in the home, rather than to become quicker and more efficient.[37] Around 1900 domestic appliances were crucial indicators of modernity, but it was the middle-class woman who spearheaded this, since she needed and could afford them, whereas 'the upper-class woman with her retinue of servants did not necessarily need the innovations, while the working-class woman could not afford them'.[38]

A Cottage Idyll

At the same time that the interior equipment of the middle-class home came to symbolize modernity, middle-class housing design developed both progressive and retrogressive features. In part it was stimulated by the reformist attitudes of Arts and Crafts practitioners, but it was also shaped by the activities of philanthropists and social reformers who idealized the 'English' domestic architectural traditions exemplified by the cottage. In contrast to the unhygienic, overcrowded slums, terraces and tenement blocks of the cities, the new suburban cottages and model estates typified progressive, modernizing tendencies in the early years of the twentieth century. They were well planned and laid-out with carefully designed interiors equipped to reasonably high standards, with modern plumbing, good lighting, heating and ventilation. Nonetheless, emotionally and visually, they harked back to notions of national identity that resided in the rural 'English' countryside. The Garden Cities and model estates incorporated the countryside into the town and city, with closes, greens, avenues and broad tree-lined grass verges. For the houses themselves, a panoply of traditional 'English' architectural features were used that originated in the architecture of the preceding 400 years, combined with enthusiasms for the Gothic Revival, the Queen Anne Movement and the Arts and Crafts Movement, as well as the ideas of John Ruskin and William Morris. Thus the planning and design of the new Garden Cities and suburbs combined 'Englishness' and modernity.

Reformers such as Rowntree had highlighted the cramped and inadequate conditions in which many people lived in his 1899 survey of York. In 1914 he estimated that 3 million people in Britain were living without minimum standards, particularly light, space, water, heat and ventilation.[39] Before this the Quaker philanthropists Lever and Cadbury (soap and chocolate respectively) had undertaken the design and building of worker housing at Port Sunlight (1889) on Merseyside and at Bourneville (1879) outside Birmingham. The housing was for their factory workers, and they incorporated features that were highly progressive: good-sized gardens, picturesque layout (in the case of Port Sunlight) and houses set back from the street. These model estates deployed a mixture of architectural idioms, although principally they combined Arts and Crafts cottage styles and Queen Anne features: small-paned, white-painted sash windows, red-brick construction with large chimneys, black-and-white half-timbering, with decorative elements that looked back to traditional vernacular and domestic styles from the fifteenth to the eighteenth centuries. A number of architects were involved with these developments, including William and Segar Owen, J. Lomax Simpson and later, Edwin Lutyens and Charles H. Reilly.

The Quaker chocolate manufacturer Joseph Rowntree and his son Seebohm drew on their first-hand knowledge of poor housing in York when they built a model

Housing at New Earswick, York, c. 1905. Large back gardens in Poplar Grove and Station Road enabled residents to grow vegetables and fruits for domestic consumption.

estate outside the city. Building on 150 acres of land near the village of Earswick, two-and-a-half miles to the north of the centre of York, Rowntree commissioned the planner and designer Raymond Unwin and the architect Barry Parker to produce an overall plan for a new 'garden' village as well as plans for housing. The housing at New Earswick was not only for workers in Rowntree's factory, but also for other working-class families in York. The rents, of between 4s. 6d. and 7s. 9d., were within the reach of those earning 26s. per week – 62 per cent of working people in York, according to Rowntree's study – and the aim was to provide each house with a garden, a view and access to the road.[40]

Parker and Unwin were in partnership from 1896 in Buxton, Derbyshire, and their ideas were shaped by Arts and Crafts thinking. Parker had trained at the South Kensington School of Art in London in 1886, and was articled to the Manchester architect G. Faulkner Armitage in 1889, who had a workshop as well as a smithy, thus providing 'an excellent training ground for a young Arts and Crafts architect'.[41] From early on, Unwin had been interested in social issues. He had attended Ruskin's lectures at Oxford and had also met William Morris. As his wife, Ethel, put it,

> Unwin had all the zeal of a social reformer with a gift for speaking and writing and was inspired by Morris, Carpenter and the early days of the Labour Movement. Parker was primarily an artist. Texture, light, shade, vistas, form and beauty were his chief concern. He wanted the home to be a setting for a life of artistic worth.[42]

In their various writings, including the book *The Art of Building a Home* (1901), they drew on Arts and Crafts Movement ideas, and their plans for model estates, Garden Cities and suburbs helped to popularize Arts and Crafts design principles.[43]

Preferring simple vernacular styles for architecture, but appalled by the monotonous rows of working-class housing found in most British towns and cities, Parker and Unwin had a strong commitment to improving the standards of housing for the working classes. The building of New Earswick was an attempt to create a balanced village community where, although rents were to be kept low, there should be a modest commercial return on the capital invested. New streets, such as Station Road, Poplar Grove and Ivy Place, were made up of groups of four to six houses; they had large gardens with fruit trees and enough ground to grow vegetables; and there was ample space to dry clothes outdoors. Generous open green space, grass verges with trees and playgrounds provided space for children, and by 1912 a co-educational primary school had been built. In 1905 the Folk Hall

Scullery with bath and cover at New Earswick, pre-1920.

Scullery showing sink in front of window at New Earswick, pre-1920.

opened as the community centre, followed by a bowling green and tennis club, and there were shops and good transport links to York. Contemporary postcards and photographs of New Earswick depict a healthy, rural, life-enhancing village where inhabitants enjoyed gardening and leisure activities, in marked contrast to the lives of York's slum dwellers. The blocks were architecturally varied and arranged picturesquely, while the cottage-style housing referenced Arts and Crafts styles exemplified by the work of C.F.A Voysey, M. H. Baillie Scott and Lutyens with gables and barge board-ing, but later housing became simpler, using a neo-Georgian vocabulary with sash windows and more regular facades.

New technologies found in these homes comprised coal-fired ranges for everyday cooking and hot water via a back boiler; coppers and lighting were fuelled by gas. Ranges were either in the centre of the main living room or in the scullery. Other novel features included inside baths and toilets and built-in cupboards that were placed in recesses and on landings for the stor-age of crockery and linen. Interiors varied according to incomes. The more modest houses had no parlour, but instead had a long living room from front to back with a window at each end to give light at all times. The par-lour, viewed by Unwin as merely imitative of middle-class values, had been observed by Rowntree in his 1899 study as being an under-used space, and, given the limitation on costs, it was not included in many of the houses. This was contentious to some extent, because working-class women were keen to have a room separate from the living room in which the family lived, ate and cooked. The parlour provided a place in which objects and furniture of value could be displayed. Additionally, it was the interface between the private world of the family and the wider public domain, a space to meet the doctor, the vicar and the undertaker. In the early houses, baths were in the scullery, covered with a wood board so as to provide a workspace when not in use, but by the mid-1920s houses had baths and toilets in a separate indoor room. The scullery had a sink with running hot and cold water, and space for preparing food under a large bright window. Photographs show tidy, well-kept interiors with an open coal fire and range, built-in cupboards, ceramic tiled floors with mats and simple furniture in the sitting room – a table and dining chairs – with one or two comfortable chairs and side tables. A profusion of ornamental ceramic vases, jugs and plates, plants and framed prints is shown in these photographs. These not only represented working-class interiors at New Earswick, but they also helped to constitute specific consumption practices and to reiterate particular working-class 'desires' for certain types of goods, services and interior designs: a copper and inside bath; a cosy fireside with comfortable chairs; ordered and clean cupboards; a kitchen with sink, light window and two taps providing hot and cold running water; and an array of decorative objects – well-polished brass fenders, coal scuttle and tools, upholstered furniture with ornamental cloths and covers, and decorative, matching jugs, vases and bowls. More than merely describing ideal working-class 'homes', these photographs choreographed working-class identities by design.

Parlour at New Earswick, pre-1920s. Displayed in this room are the 'best' things: upholstered chairs, potted plants, mantelpiece and mirror, decorative ceramics and family photographs.

Subsequent historians have noted that, for all their radical planning, the cottage designs of Parker and Unwin perfectly expressed 'the ideology of women as keepers of the domestic sphere'; situated well away from industry, they inscribed the separation of men's paid work outside the home from women's unpaid labour within it.[44] Initiated in a context of enlightened patronage that extended philanthropic traditions, the design and development of New Earswick was also prompted by Rowntree's ground-breaking social survey of poverty in York of 1899. To an extent paternalistic, Rowntree nevertheless observed the problems faced by working-class women, and the design of New Earswick marked an attempt to overcome this. He wrote:

> No one can fail to be struck by the monotony which characterises the life of most married women in the working class. Probably this monotony is least marked in the slum districts, where life is lived in common, and where the women are constantly in and out of each others' houses, or meet and gossip in the courts and streets. But with the advance of the social scale, family life becomes more private, and the women, left in the house all day whilst their husbands are at work, are largely thrown upon their own resources. These, as a rule, are sadly limited, and in the deadening monotony of their lives these women too often become mere hopeless drudges.[45]

The architects Unwin and Parker tried to address this problem in the overall layout of New Earswick. They allowed for communality by placing streets at right-angles and in small closes, and by putting large rear gardens back-to-back so as to enable informal social interaction. Retrospectively, it is clear that New Earswick, along with Letchworth, contributed to the development of what was to be the preferred model of cottage estates for state-funded working-class housing both up to and beyond the Second World War.

As with New Earswick, the development of the first Garden City at Letchworth in 1902 encompassed both progressive and reactionary tendencies. Parker and Unwin were the architects/planners of this first practical demonstration of Ebenezer Howard's ideas as outlined in his book of 1898, *Tomorrow: A Peaceful Path to Reform*, reissued as *Garden Cities of Tomorrow* in 1902. Howard called for 'an earnest attempt . . . to organize a migratory movement of population from our overcrowded centres to sparsely settled rural districts'.[46] As a means to address urban overcrowding, the Garden Cities were essentially designed to create a 'magnet' to draw people back into the countryside, offering 'all the advantages of the most energetic and active town life, with all the beauty and delight of the country'.[47] Between 1891 and 1911 there was a significant move, nationally, outwards to the suburbs. London led the way, attracting people to the suburbs from rural areas as well as from metropolitan centres, while the population of inner London remained static.[48] In order to manage this, a Garden City was planned on a site just over 30 miles north of London at Letchworth. The aim was to provide decent, hygienic housing at low rents, although rents of 5s. 6d. were beyond the reach of the poor who had been provided for at New Earswick.

Adapting the progressive elements of the late nineteenth-century domestic revival to the needs of an essentially urban population, Parker and Unwin designed housing that employed English vernacular styles to meet twentieth-century housing needs via a plan that included tree-lined culs-de-sac, greens and avenues, and broad grass verges combined with short terraces and semi-detached blocks of low-density housing.[49] Interior design was carefully thought out, including built-in cupboards and other furniture, such as inglenook seats placed alongside the range. All housing was separated from industry by strips of planted woodland; it was set back and included ample gardens, hedges and footpaths, and although the Arts and Crafts vernacular exteriors looked back to the past, the modernity of Letchworth was evident in terms of town planning and interior design. The Garden City approach to urban planning and house design was perceived at the time to represent a unique 'English sensibility'. This was apparent to Hermann Muthesius, a German architect attached to the German Embassy in London, who, in his extraordinary account of English domestic architecture at the turn of the century, *Das Englische Haus*, wrote:

> The Englishman builds his house for himself alone. He feels no urge to impress, has no thought of festive occasions or banquets and the idea of shining in the eyes of the world through lavishness in and of his house simply does not occur to him. Indeed, he even avoids attracting attention to his house by means of striking design or architectonic extravagance, just as he would be loth to appear personally eccentric by wearing a fantastic suit. In particular, the architectonic ostentation, the creation of 'architecture' and 'style' to which we in Germany are still prone, is no longer to be found in England. It is most instructive to note . . . that a movement opposing the imitation of styles and seeking closer ties with simple rural buildings, which began over forty years ago, has had the most gratifying results.[50]

Admiring the lack of architectural ostentation typified by the English domestic architecture, Muthesius also noted its lack of 'style' and its modest individuality. *Das Englische Haus* highlighted the details and practical qualities of these buildings as well as the reformist impetus of their architects. It is telling that these highly modern, socially progressive ideals, which led to a fundamental shift in thinking about housing the urban working class, summed up by New Earswick and Letchworth, should be 'clothed' in visual metaphors that evoked a stable rural 'Englishness' prior to the massive industrial change of the late seventeenth and eighteenth centuries. This vision of Englishness ameliorated the radical edge of modernity. These contradictions, which marked out the design and planning of the model estates such as New Earswick and the Garden Cities such as Letchworth, were the very same qualities that attracted the interest of socially reformist modernist architects in the early years of the twentieth century.

Arts and Crafts North of the Border

While the deployment of Arts and Crafts Movement strategies in 1890 represented a radical stance with social and political implications, by 1914 it had become fractured into a number of different, sometimes contradictory elements. It still remained a powerful visual language that was utilized by manufacturers and advertisers to suggest authenticity and continuity. As has been seen, this was apparent in housing design, but it was also the case in a variety of manufacturing industries as demand for a handmade aesthetic increased. Josiah Wedgwood and Sons Ltd, the prestigious ceramic manufacturers, employed Arts and Crafts designers such as Alfred and Louise Powell, and Grace Barnsley, daughter of the Arts and Crafts furniture maker Sidney Barnsley, to produce designs for mass-production. But these were nevertheless hand-painted using colours and patterns influenced by the work of Arts and Crafts designers such as William De Morgan, C.F.A. Voysey and C. R. Ashbee. These designs represented the commercialization of the Arts and Crafts Movement, even though the quality was high – both aesthetically and technically.

For many committed to craft aesthetics, the ideas of Ruskin and Morris lost their appeal, and they focused their attention on materials, processes and techniques, rather than wider social issues. A new generation of craftsmen and women emerged whose work enhanced and transformed the practices of craft. Potters such as Reginald Wells were typical, and the emergence of studio pottery in the 1910s, for example, marked a sea change in thinking about the craft of ceramics. Studio pottery was developed by younger potters who were neither ideologically linked to the Arts and Crafts Movement nor practically predisposed towards art pottery made within an industrial context. Instead, they were searching for a new approach to handmade ceramics. Like many involved in the Arts and Crafts Movement, Reginald Wells was well informed about early English pottery traditions such as slipware, but whilst studying at the Camberwell School of Arts and Crafts he developed an interest in early Chinese ceramics, leading him to experiment with glazes. His pottery, described as 'unlike that of any other modern potter', relied for its effect on relatively simple shapes combined with stunning colour and surface effects.[51]

There were, however, many who remained loyal to Ruskinian beliefs. Significantly, a large number of these were women who contributed to a renaissance of craft ideas from the 1910s through to the '30s.[52] Art schools provided crucial training and the curricula were imbued with Arts and Crafts Movement values and practices.

Embroidered cushion designed by Ann Macbeth, sewn by Gertrude Young and Kate Catteral, 1906.

Glasgow School of Art was well known for this and the work of the embroidery class established in 1894 was exemplary. Led by Jessie Newberry, a creative and inspirational needleworker and the wife of Francis Newberry, the school's embroideries used simple as well as rich materials, such as unbleached calico, linen and flannel, crewel wools and silk threads, in addition to less orthodox techniques such as appliqué and needle-weaving. By 1902 *The Studio* magazine wrote: 'Look to the Glasgow School of Art if we wish to think of today's embroidery as a thing that lives and grows and is therefore of greater value and interest than a display of archaeology in patterns and stitches.'[53] Offering a certificate in Art Needlework and Embroidery from 1907, the embroidery classes made a significant contribution to the craft of needlework. In addition, books such as *Educational Needlecraft* written by Ann Macbeth (who succeeded Newberry as head of Needlework at Glasgow) and Margaret Swanson in 1911, as well as the special Saturday morning classes for teachers, perpetuated an Arts and Crafts approach beyond the First World War.

In Edinburgh, 'the beginnings of Arts and Crafts practice lay in philanthropy'.[54] A good example was the Edinburgh Social Union, which included social thinkers such as Patrick Geddes and designers such as Phoebe Anna Traquair. From the late 1880s Traquair's interest in embroidery, mural, bookbinding and illumination led to the development of her craft skills. Supported by the Edinburgh Social Union, her work frequently had a philanthropic dimension, since she designed decorative schemes for churches, halls and hospitals. In direct correspondence with Ruskin in the 1880s

Designing Modern Britain

Murals in the chancel arch of the Catholic Apostolic Church, Mansfield Place, Edinburgh, 1893–5, by Phoebe Anna Traquair.

regarding bookbinding, Traquair was already committed to the principles of design that he and Morris had outlined. Her own decorative work incorporated figures influenced by the paintings of Rossetti and Burne-Jones as well as Renaissance art, but they also hint at modernity through the use of abstracted forms, unusual materials and techniques, and bold colour (as seen in the murals of 1893–5 in the Catholic Apostolic Church in Edinburgh).

By the 1890s Edinburgh was a centre for the Arts and Crafts Movement in Scotland, attracting artists, designers and thinkers from England and Scotland, including William Morris, Walter Crane and Francis Newberry. In 1892 the School of Applied Art and the Old Edinburgh School of Art were established, and along with organizations such as the Edinburgh Social Union, these were committed to improving building, design and craft. The Edinburgh Social Union, where Traquair taught, aimed to teach people

> to beautify the homes of the poor but [also] to teach them to do so themselves . . . to establish a little artistic society, where workers can come and try to work out for themselves the answer to some of the difficulties in the way of getting beauty introduced into the common objects of everyday life . . . [55]

But increasingly the work of some Scottish designers began to explore themes that appeared at odds with the distinctive ethical stance adopted by the Arts and Crafts Movement in England, and this was particularly apparent after the work of the Glasgow Four (Charles Rennie Mackintosh, Herbert McNair and Margaret and Frances Macdonald) was disparaged in the English art press by Arts and Crafts supporters.

Edinburgh and Glasgow were vibrant artistic centres in the 1890s and 1900s with strong links to Europe. Edinburgh hosted an International Exhibition of Science, Art and Industry in 1886 and again in 1890 (the latter celebrated the opening of the Forth Rail Bridge in 1889). Glasgow staged the International Exhibition of Science, Art and Industry in 1888, the Glasgow International Exhibition in 1901, and the Scottish Exhibition of National History, Art and Industry in 1911. At the 1901 exhibition, the most progressive work of designer/craftsmen and women was displayed: C. R. Mackintosh designed a stall and Jessie Newbery exhibited an embroidered curtain, which, according to *The Studio*, commanded 'the highest admiration for the skilful arrangement of intricate lines, combined with workmanship of rare excellence'.[56]

Traquair had exhibited at the Exposition Universelle in Paris in 1900 and Newbery at the Turin International Exhibition of Decorative Art in 1902. Modern Scottish painting had attracted critical attention outside Britain since the mid-1890s, when work by the 'Glasgow Boys' was shown in Munich. But within ten years, modern Scottish design became equally if not more prestigious in Europe following the widespread admiration in Austria and Germany for the Glasgow Four — and the exhibition of their work at the various Secessionist exhibitions in 1898, 1900 and 1902. Enthusiasm for the work of these Scottish designers in Europe can be attributed to a large extent to its modernity and its reworking, if not rejection, of a regional or indeed a national aesthetic. Although Mackintosh's Hill House of 1902 at Helensburgh had used vernacular forms, techniques and materials, he quickly moved beyond this, and his design for the Glasgow School of Art

Designing Modern Britain

pointed to a more universal modern aesthetic linked to developments in Europe rather than reiterating the local and regional vernacular styles of architecture and design so beloved of the predominantly London-based Arts and Crafts Movement writers, educators and practitioners.

There were other serious efforts to apply Arts and Crafts thinking to the modern world in the activities of those associated with the Design and Industries Association (DIA), formed in 1915. This group of thinkers, educators and designers were committed to the design of simple, solidly made objects, but under the influence of the German Werkbund they began to conceive the idea of an aesthetic based on mass-production rather than hand-methods. In grappling with this, the DIA recognized the need to move beyond the narrow world of craft guilds and workshops in order to face up to the very pressing concerns of industry within a world marketplace. These issues came to the forefront with some urgency after the First World War.

Cultures of Shopping

Between 1897 (the Diamond Jubilee of Queen Victoria) and 1911 (the coronation of her grandson, George V), London was transformed from a medieval and haphazard city into an 'imperial' one. It was 'a city of new processional streets, a city of new palaces of gleaming white stone – palaces of banking and commerce in the West End and the City; palaces for new bureaucrats in a completed wide Whitehall'.[57] The visual characteristics of many of these buildings were ostentation and display, achieved through a plethora of historical styles. Classicism was deployed to evoke the grandeur, status and stability of 'British' imperial power. Like the dichotomies evident in the interior and exterior of the Garden City cottage, however, the internal structure and equipment of these buildings could be highly modern. The Ritz Hotel, for example, had the first internal steel frame in London, and the Hotel Russell (1898) on Russell Square had adjacent, separate bathrooms for each bedroom.[58] At its centre, London offered a magnificent setting for the production, circulation and consumption of design, and a wealth of grandiose new buildings contributed to this. Public buildings were erected or remodelled (Admiralty Arch, 1910–11); new streets were developed and improved (Kingsway, Aldwych and Strand, 1892–1905); new public monuments and statues were erected (*Eros* in Piccadilly Circus, 1892); and art galleries, museums and public institutions were built (the Edward VII Galleries at the British Museum, 1904–14; Royal Automobile Club in Pall Mall, 1908–11). New educational institutions were conceived and built (Goldsmith's College, 1907; Central School of Arts and Crafts on Southampton Row, 1907–9), and large offices were erected (Country Life

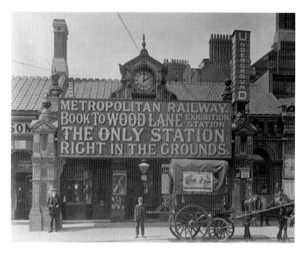

Building in Tavistock Street, 1904–6; Kodak House on Kingsway, 1911). To provide accommodation for London's myriad visitors, prestigious hotels were built (Claridges on Brook Street, 1898; the Savoy, 1903–4), and contributing to the image of London as a site of leisure, new theatres were conceived (the London Coliseum, the London Hippodrome, the London Palladium, all from around 1900).[59]

Closely connected to this was the rapid expansion of London as a retailing centre. New railway and underground systems, particularly the Metropolitan Line, gave easy access to the new or remodelled department stores: Debenham and Freebody in Wigmore Street (1907–8); Harrod's on Brompton Road (1901–5); Selfridges (1907–9) and Waring and Gillow (1906) on Oxford Street; and Whiteley's on Queensway in Bayswater (1908–12). Shopping was concentrated on the main thoroughfares: the Strand, Piccadilly and Oxford Street, and, focused on the new department stores; its principal customers were women.[60] Foremost in several respects was Selfridges, opened in 1909. Gordon Selfridge, 'one of a cohort of extremely wealthy and powerful Americans who saw Edwardian London as profitable territory', aimed to sell to the masses.[61] The interior design and planning of his new store was innovative, with nine passenger lifts, electric lighting and a steel frame, whilst the exterior was in a simplified, though monumental classical style with huge plate-glass windows. Around the country, in large cities, department stores such as Fenwick and Bainbridge (Northumberland

The Royal Coach at London's new Admiralty Arch (designed by Sir Aston Webb), 1911.

Finchley Road station, Metropolitan underground railway line, 1910.

Designing Modern Britain

Selfridges department store, Oxford Street, London, 1909. Department store architects deployed varied, but predominantly classical, styles to represent the commercial ambitions of prestigious stores located on main city thoroughfares.

Jenner's department store, Princes Street, Edinburgh, 1895. Architect, William Beattie, photographer, Henry Bedford Lemarr.

Street and Market Street, Newcastle upon Tyne) and Jenners (Princes Street, Edinburgh) extended, rebuilt and modernized in order to woo women shoppers. Like their London counterparts, these provincial stores were architecturally lavish and technologically advanced. Jenners in Edinburgh (by 1890 the largest retail establishment in Scotland) reopened following a fire in 1895 to a design by the architect William Hamilton Beattie; it included air-conditioning, hydraulic lifts and electric lighting. The exterior style, a variation on Renaissance classicism, comprised polished granite, marble, carving and statuary. Internally, the store had a main avenue paved with marble mosaics. A large central treble-height space was surrounded with galleries on the first and second floors. Painted cream and gold, the interior, including the grand staircase, was panelled with expensive woods: ebony, mahogany and oak, and the entire space was lit with copious numbers of electric fittings. Promotional literature to celebrate its reopening in 1895 noted that 'the Luncheon and Tea-rooms constitute a part of the Establishment which one lingers in with much pleasure, on account of the thoroughly artistic and elegant style in which the arrangement and decoration have been conceived'.[62]

In the second half of the nineteenth century Newcastle witnessed 'an all-time record in the city's geographical and population expansion'.[63] Large industries such as R. & W. Hawthorn, W. G. Armstrong and R. Stephenson had created one of 'the most important centre[s] in the world for the manufacture of ships, armaments and locomotives'.[64] Enterprise characterized Newcastle. This was evidenced by the works where Stephenson's famous Locomotion and Rocket railway engines were made by 'a factory producing in the 1860s the most accurate rifled guns in the world; the first factory to produce electric lamps on a commercial scale (1881); and another making the world's first turbine-powered (alternating current) power stations (1892)'.[65] Newcastle's wealthy classes benefited from all this, building or remodelling large country houses away from the city from the mid-nineteenth century onwards. They employed well-known architects such as R. Norman Shaw, who built Cragside in Northumberland in the 1870s and '80s for Armstrong and remodelled Chester in the 1890s for the Claytons. These families were also patrons and participants in Newcastle's cultural, social and intellectual life, contributing to the Literary and Philosophical Society and establishing the College of Physical Science (later Armstrong College, part of Newcastle University).

The city of Newcastle upon Tyne was also the premier site in the northeast of England for shops and shopping. Influenced by the building of the Burlington and Lowther Arcades in London, Richard Grainger, the Newcastle builder and developer, had undertaken the building of the Royal

Designing Modern Britain

Arcade in 1831–2. This included 16 shops, banks, auction rooms, offices, a post office and 'a steam and vapour bath'.[66] From the 1830s and '40s onwards Grainger undertook further building and development in Newcastle, and streets such as Grey Street, Grainger Street, Market Street and Clayton Street were laid down, creating 'the best designed city in England'.[67] The architects included John Dobson, John Walker and George Wardle, the latter responsible for the commercial buildings on Market Street, which were occupied by Bainbridge department store from the 1840s.

Bainbridge was organized by departments as early as 1849, making it one of the earliest department stores in Europe.[68] By 1899 it had extended along Market Street and into the Bigg Market, employing 700 staff (many of whom lived in). New technologies were introduced into the store at the end of the nineteenth century, including gas-engine-powered cash tubes, a cable tramway for moving goods from the wholesale packing area, a carpet-beating machine and, in 1902, a Panhard Lavassor motor delivery van bought in Paris.[69] In 1911 an impressive new shop frontage was opened with arcade windows, and in 1913 further extensions into the Bigg Market, which included a new tea room with Tudor-inspired oak panelled walls and ceiling. In common with other department stores, Bainbridge developed innovative promotional strategies to attract and keep its customers. Described as 'The Emporium of the North', a brochure in 1913 advised:

> For the Lady unable, on occasion, to shop in person, there is our Telephone Service, which puts her into direct communication with *any Section or Counter in our warehouse*. This ensures the careful fulfilment of her instructions, and prompt delivery of the goods.[70]

The brochure listed 13 departments: dress materials, silks, Ladies' Wear, fancy, Gents' Wear, boots, drapery, flannels, cabinet furniture, soft furnishings, floor coverings, linens and hardware, and included a photograph showing fashionable shoppers admiring the new window displays. A company calendar of 1914 showed the interior layout and design of a number of departments – dress materials, drapery, blouses and ladies' wear. These were spacious and opulently decorated with luxury carpets, fretted woods, moulded and ornamented friezes, extensive glass-fronted cabinets and electric lighting.[71] Not only were these department stores opulent and plush – unlike most women's homes – but they also represented

an exciting adventure in the phantasmagoria of the urban landscape. The department store was an anonymous yet acceptable public space and it opened up for women a range of new opportunities and pleasures – for independence, fantasy, unsupervised social encounters, even transgression.[72]

Newcastle's other main department store was Fenwick on Northumberland Street. J. J. Fenwick had opened a small mantle and furrier's shop there in 1882, followed by a store in Bond Street in London in 1891. Directly influenced by Parisian stores such as Bon Marché, Fenwick opened a new store on Northumberland Street in 1913 designed by the architects Marshall and Tweedy in an Italian Renaissance style. The local press described its opening in October 1913 as the 'Business Triumph of Progressive Newcastle Firm', and photographs showed some of the 20,000 people – who passed through on its opening day – queuing to get into the north east region's premier department store.[73] Housing 22 departments, it included six 'exclusive ones'; these were dressmaking, furs, millinery, young ladies' costumes, mantles and tailor gowns. Typically luxurious and clearly aimed at the woman shopper, the materials used for the interior were lavish, echoing the interior schemes of the private houses of the fashionable and wealthy: traditional 'English' oak and mahogany panelling, light fawn hand-tufted Donegal carpets, oxidized silver electric candelabra and mosaic panels. Women shoppers were free to wander and indulge in the visual pleasure and spectacle of the interior and the goods on display. By 1913 Fenwick employed 200 staff in its Newcastle shop and 100 in its London store. In sale catalogues from June 1909 advertising original Paris costumes by Worth, Bernard and Cheruit, the dressmaking department offered a 'model gown in sulphur coloured faced cloth, with bodice of crepe-de-chine to match, handsomely trimmed' for $7\frac{1}{2}$ guineas instead of 14 guineas.[74] To transport women to the sale, convenient trains were listed from Alnmouth, Barnard Castle, Carlisle, Corbridge, Darlington, Durham, Harrogate, York, Middlesbrough, Rothbury, Saltburn and Sunderland. An obituary of the founder, John James Fenwick, in the *Northern Echo* on 12 July 1905, described him as 'the Worth of the North', and it was said that his designs were 'at once strikingly original and fully in keeping with the most cultured taste of the period'.[75] He made his reputation supplying 'every well-known family in the North, and society women in the West-End of London with rich furs'.[76] But at the same time, Fenwick also offered an 'inexpensive, smart coat and skirt in New Tweed, cut and fitted by men, sewn by tailoresses. Entirely made on the Premises in healthy workshops.'[77]

Designing Modern Britain

Displays of Power

At first glance women's fashionable dress was both steadfastly 'imperial' and shockingly sexualized. The epitome of power, the extravagantly dressed wives and daughters of the ruling classes, deployed fashion with some aplomb to convey their social superiority. Elaborate coiffures, delicate lace and embroideries and tight-laced Parisian gowns combined to produce an exaggerated, highly sexualized femininity that seemed the antithesis of modernity.[78] But by the end of the nineteenth century and the beginning of the twentieth, the modern city was increasingly peopled by the working classes as well as the wealthier ones. In its public spaces, female identities were being transformed, not just politically via the suffrage campaigns, but also by an array of artistic, cultural and social representations.

The changes that women's work underwent between 1890 and 1914 were caused in particular by the impact of new technologies, which enabled women to make inroads into previously male-defined categories of employment. By the 1890s women were challenging patriarchal assumptions about appropriate work by taking on jobs previously done by men. Initially castigated as 'unwomanly' and criticized for inflicting untold injuries on themselves and society, by the early years of the twentieth century these women were earning better wages, and, after considerable struggles, they had joined trade unions and had begun to reap the benefit.[79] Photographs and early promotional films from these years show young women factory, office and shop workers heading off to work wearing fashionable, but more practical clothes. A photograph from the 1890s depicting a throng of workers arriving at the Army and Navy Stores reveals women wearing shortened flared skirts and 'mannish' jackets, shirtwaist bodices (sometimes with bow ties) and straw hats.[80] These seem a world apart from well-dressed metropolitan socialites, but these women were part of related structures of modernity that connected the economic, political and cultural landscape of Britain. They were visible and on display in the streets of large cities and towns. Working, shopping and meeting, as well as attending art galleries and museums, the theatre and restaurants, art school and university, they were essential components of modernity. Fashion both represented and encouraged all this; it was the embodiment of modernity and integral to and part of modern capitalism. Women from the upper and middle classes in particular formed a huge potential market eager for an array of goods whose aim was to enhance and beautify the female

Fashion illustration from *The Lady's Realm*, Henry Bedford Lemere, 1906.

body that was displayed in department stores and drapers' store advertising and discussed in women's magazines.

Cultures of public display were a defining feature of Edwardian Britain, particularly those evident in the large exhibitions organized during the period. Motivated in part by a desire to engender national unity, the Festival of Empire exhibition held at Crystal Palace in 1911 was also a symbol of the economic imperatives of high capitalism in its promotion of British (rather than English or Scottish) products and foreign trade.[81] Its model was the large number of international shows organized following the Great Exhibition in London in 1851, especially the Columbia World Fair in Chicago in 1893, which had the world's largest ferris wheel at 250 feet high.

Funded by businessmen and politicians rather than central government, these exhibitions aimed to consolidate, articulate and render 'natural' the unity of the nation and the empire.[82] The organizing council of the Festival of Empire numbered among its members the requisite royalty, HRH Princes Louise, Duchess of Argyll (Honorary President of the Ladies' Committee), as well as the Duke of Devonshire, who was President of the British Empire League; the Earl of Plymouth, Chairman of the Council; the artist Sir Edward Poynter, President of the Royal Academy; the architect Sir Aston Webb, President of the Amphitheatre Committee; and the Master of the Festival, Frank Lascelles. The aim of this exhibition was 'to demonstrate to the somewhat casual, oftentimes inobservant, British public the real significance of our great Self-governing Dominions – to make us familiar

Designing Modern Britain

with their products, their ever-increasing resources, their illimitable possibilities'.[83] It was planned for the year of George v's coronation, entirely appropriately, it was felt, since he was 'the first British Monarch who has both visited and made himself acquainted with the outposts of the Empire'.[84] The Crystal Palace was used to display the goods of the empire, with special sections devoted to applied chemistry, engineering, mining, imperial industries, art, crafts and home industries, decoration and furnishings, perfumes, textiles, shipping, sports and pastimes. The extensive grounds were formally arranged around a central 'Empire Avenue' and 'Grand Parade' with a spectacular amphitheatre designed by Aston Webb, using the imperial vocabulary of the Colosseum in Rome. The dominions were represented with three-quarter-size replicas of their parliament buildings, and there were films, models and pictures of colonial life in the interiors. This was set against a backdrop of painted scenes of typical landscapes from around the world, and played out against this was the Great Pageant organized by Frank Lascelles. The pageant, described as 'a new popular movement which had its birth but a few years ago, and which has steadily been spreading over rural Britain', and 'a symptom of developing taste among the people', was achieved by means of a sequence of screens depicting the history and development of the empire with scenes acted out by 15,000 performers.[85] It started with a scene from the 'Dawn of History' progressing through prehistory to the 'Age of Chivalry', followed by 'Social Upheaval' (Wat Tyler of the Peasants' Revolt) to 'Elizabeth and the Return of Drake', on to 'Trade with the Indies' and culminating in the 'Grand Imperial Finale', via Australia, New Zealand, South Africa, Canada and India.

The deployment of this particular historical narrative to symbolize the continuity, permanence and overriding logic of white imperial Britain was highly revealing during a period of intense debate about the future of the empire. Not only was there government concern that the empire at this time was a 'gigantic global juggernaut, spinning quite out of control', but there was also anxiety about the 'swamping' of the constitution by the masses.[86] In such a context, the spectacle and drama of the Festival of Empire and Imperial Exhibition of 1911 – particularly the Great Pageant – represented a form of visual reassurance. Essentially an urban drama, its site at Crystal Palace embodied the 'charm of rural England, and yet it is sufficiently near the Empire City for the throb of the great heart of London to be audible'.[87] Evoking an 'Englishness' based on tradition and stability, and a 'Britishness' that was both imperial and modern, it also pointed to the contradictory experience of modernity. As the Festival of Empire's Great Pageant constructed a tale of progress, order and continuity, public display and spectacle were marshalled on the streets of London and Britain's large

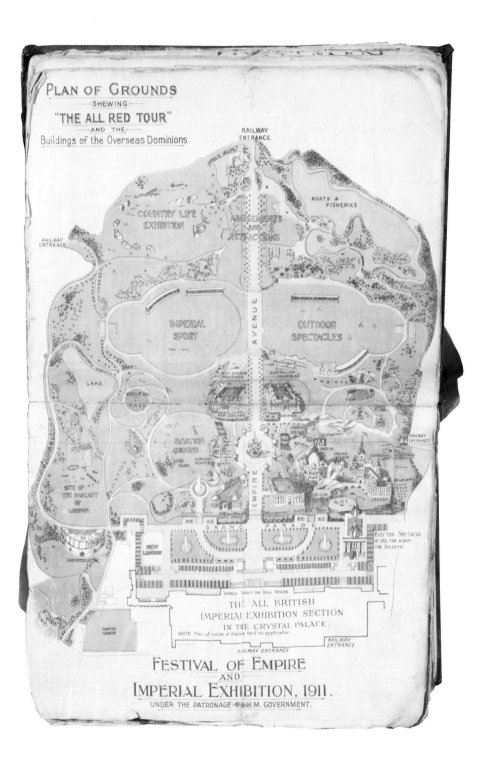

PLAN OF GROUNDS
SHEWING
"THE ALL RED TOUR"
AND THE
Buildings of the Overseas Dominions

RAILWAY ENTRANCE

COUNTRY LIFE EXHIBITION

AMUSEMENTS AND ATTRACTIONS

BOATS & FISHERIES

RAILWAY ENTRANCE

IMPERIAL SPORT

A V E N U E

OUTDOOR SPECTACLES

LAKE

MEDIÆVAL MAZE

NEW ZEALAND INDUSTRIES

AUSTRALIAN INDUSTRIES

NEW ZEALAND

AVIATION GROUND

AERO CLUB

AFRICAN INDUSTRIES

CROWN COLONIES

INDIA

CANADIAN INDUSTRIES

RAILWAY ENTRANCE

EMPIRE

SITE OF THE PAGEANT OF LONDON

CANADA

SOUTH AFRICA

NEWFOUNDLAND

GRAND PARADE

ELECTRIC SPECTACLE OF SEA TRIP ACROSS THE ATLANTIC

NEW LONDON

IMPERIAL TERRACE AND ROYAL PAVILION

THE ALL BRITISH
IMPERIAL EXHIBITION SECTION
IN THE CRYSTAL PALACE.
NOTE: Plan of inside of Palace sent on application

RAILWAY ENTRANCE

RAILWAY ENTRANCE

FESTIVAL OF EMPIRE
AND
IMPERIAL EXHIBITION, 1911.
UNDER THE PATRONAGE OF H.M. GOVERNMENT.

cities with quite different ends in view: female suffrage. Drawing, to a certain extent, on the language of trade unionism, but also the visual strategies of the pageant, these feminist visual displays were part of a design culture of resistance and politics.

At a meeting of the Women's Suffrage Societies at the Albert Hall in June 1908, Sylvia Pankhurst described the

> striking pageant with its many gorgeous banners, richly embroidered and fashioned velvets, silks and every kind of beautiful material and the small bannerettes serving as innumerable patches of brilliant and lovely colour, each one varying both in shape and hue.[88]

This pageant articulated a radically different history, not British, imperial or masculine. Instead, Joan of Arc, Elizabeth Fry and Mary Wollstonecraft were part of a procession that 'was acknowledged to be the most picturesque and effective political pageant that had ever been seen in this country'.[89] The suffrage movement's appropriation of the pageant represented shrewd tactics. In adopting this device, the suffrage societies drew on a language already well understood by the general public. Pageants were based on popular representations of the past, but the suffrage societies used them as a means of resistance, drawing on a shared sense of history, but subtly undermining it. Elizabeth I was part of the 1908 pageant at the Albert Hall, but instead of appearing alongside Francis Drake as she did in the Festival of Empire, she was part of a procession of 'well-known' women. Equally, the Pageant of Women's Trades and Professions held in London on 27 April 1909 saw 1,000 women from a wide range of jobs and professions, as diverse as doctors, pitbrow lasses, charwomen, artists and craftswomen, tailoresses, machinists and teachers. The involvement of women workers undercut the Edwardian preoccupation with order, hierarchies and continuity.[90] The pageant, in this instance, had a double political role: to represent women's coming together to demand equal suffrage, but also to show the differences between women.

To coincide with the coronation of George V on 22 June 1911 (the event celebrated by the Great Pageant at the Festival of Empire), the Women's Social and Political Union with assistance from the National Union of Women's Suffrage Societies organized The Women's Coronation Procession on 17 June 1911. This rivalled the official celebrations and was the high point of suffrage public display, aiming to persuade the vast number of people in London for the coronation celebrations of 'the scope and strength of British women's demands for the same right now enjoyed by women in several of the colonies'.[91] Pageants were designed as part of this and included

Festival of Empire exhibition map, 1911. The Master of Festival was Frank Lascelles and the architect of the amphitheatre, Sir Aston Webb.

the Historical Pageant, the Pageant of Empire and the Prisoners' Pageant. The last comprised 700 women who had been arrested or jailed during the suffrage campaigns. Significantly, women were also important contributors to the design and production of these historical pageants. Miss M. P. Noel, honorary secretary of the Great Pageant at the Festival of Empire in 1911, performed a number of roles in the pageant, as well as making various costumes, including a James II townswoman, a

The Prisoners' Pageant at the Women's Coronation Procession, 1911.

medieval damsel, a Georgian lady and, on one occasion, a beggar.[92] Equally, the various suffrage societies were dependent on the skills of their numerous members. Edith Craig, daughter of the Queen Anne architect E. W. Godwin and the actress Ellen Terry, was one such person, along with the stained-glass designer Mary Lowndes. With a background in the Arts and Crafts Movement, Lowndes was predisposed to representations of medievalism and chivalry, but, as we have already seen, this visual language of 'Englishness' had particular potency at the time. It was more significant that numerous women could contribute their design skills as needlewomen and seamstresses to the pageant displays and costumes irrespective of their formal artistic status. In this respect, the pageant was as disruptive of artistic hierarchies as it was of social ones. As an example of design culture, the suffrage pageants shared some of the qualities of other Edwardian pageants: it was a mass activity (the Festival of Empire's Great Pageant had 15,000 performers and The Women's Coronation Procession had 40,000 participants), drawing on popular historical representations. In fundamental ways, however, the design of the suffrage pageants and displays transgressed the rules. They were overtly political in that design and material culture was used for political ends in the campaign for women's suffrage, an issue that undermined any pretence of stability and continuity; they disrupted social hierarchies and class divisions by making visible wider women's participation, and they undercut hierarchies of artistic value by deploying design in its broadest definitions.

Ultimately, design in late Victorian and Edwardian Britain was paradoxical. At the same time that it conjured up particular images of 'Englishness', 'Britishness' and 'Empire', it also represented the transformative experiences of modernity. As being 'British' became something of a moveable feast within a changing empire, it was significant that Scottish design looked eastwards to Europe, particularly to Austria and Germany. With less of an incessant pull towards tradition, Scottish design intimated at modernity and internationalism. In contrast, the deployment of formal

Designing Modern Britain

classical styles to symbolize the power of empire, summed up by the rebuilding of London, department store design, and national and international exhibitions, highlighted Britain's economic authority and pointed to the importance of the consumer within the culture of high capitalism. Parallel to this, however, design languages rooted in the Elizabethan and Jacobean periods were also used for interior design and fittings in large department stores, representing a high point of a specifically 'English' culture. As Edwardian upper-class women's fashion was sexual, imperial and stridently elite, suffrage displays and demonstrations from 1907 up to the outbreak of war in 1914 used design subversively to demand the vote. But at the same time, women's fashions were genuinely transformative and the practical stylish jacket and skirts worn by the Army and Navy workers demonstrated the extent to which everyday women's engagement with contemporary fashion was also an engagement with modernity. Inevitably 'Englishness' and 'Britishness' were precarious and difficult to define – to a certain extent the product of a hugely volatile era prior to the First World War. British design both contributed to and responded to this, and retailers, advertisers and consumers were simultaneously intrigued by both urban and rural, by tradition and modernity, but also by progress, rationalism and the promise of the future. As chapter Two shows, designers and manufacturers looked back to seventeenth- and eighteenth-century English traditions, forward to modernism and mass consumption, and eastwards to non-Western traditions after the First World War.

2

'Englishness' and Identity: Design in Early Twentieth-century Britain

The loss of export markets and economic collapse in the 1920s, the reshaping of empire and Britain's relationship to it, and changing class and gender identities all contributed to a realignment and fragmentation of what constituted 'Britishness' in the early part of the twentieth century. The possibility of mass democracy came with the Representation of the People Act of 1918, which trebled the electorate by giving all men over 21 and women over 30 the vote (all women gained the vote in 1928). It was still the case that the people of England were predominantly 'white', but as the British Empire expanded after the League of Nations settlement at the end of the First World War, Great Britain most assuredly was not. Empire included not only the mainly 'white' British Isles and its Dominions – New Zealand, Australia and Canada – but the subcontinent of India, large areas of Africa and the Middle East (existing colonies such as Kenya and Nigeria were joined by new territories such as Tanganyika, the Cameroons, Iraq, Jordan and Palestine).[1] In such a context 'Britishness' was increasingly permeable, providing the ideological lynchpin of a huge new trade configuration with 500 million potential 'consumers' – enough to rival the population of the USA (130 million) and Europe (450 million).[2] If 'Britishness' represented expansionism, then Englishness, by contrast, 'stood for the private and intimate, the spiritual and the primitive', located in the countryside rather than the city.[3]

Change and continuity went hand in hand. Magazines such as the *Burlington Magazine*, *Vogue* and *Country Life*, for example, simultaneously evoked tradition and modernity by referencing the metropolitan and the rural in design. The product of refined taste and connoisseurship, of a country-house culture of elegant women, weekend parties and interior decorating and style, design as represented in *Vogue* and *Country Life* was

Hand-woven scarf by Ethel Mairet, 1933–5. Black, brown and beige machine-spun cotton warp, machine-spun undyed wool and single tussah silk weft.

47

also connected to the glamour of the modern city: London, Paris and New York. London was at the centre of modernity and at the confluence of different global cultures, from America, Russia and the British Empire. Modernity spread out to the regions, too, by a variety of means – the new multiples, Woolworth and Marks and Spencer, advertising, popular music and dancing, and, of course, the cinema – the new media of the 1920s. An emphasis on the 'visual' was an important feature of 1920s modernity combined with a sense that life had speeded up via the new Underground, the motor car and the vast ocean liners. Even the dances were quicker and more energetic, particularly the Charleston arriving from America in 1926. America was the source of much of this popular modernity, but it attracted criticism as 'low brow'. Jazz was a case in point, prompting the Bloomsbury art critic Clive Bell to declare that it stood for 'the chaos of the mind'.[4] Everyday goods and advertising were particularly susceptible to American influences with 'jazz' motifs – angularity, bold colours and bold pattern affecting an array of things. But in design terms 'Englishness' remained important. Encapsulated by a number of historical revivals and an ongoing concern for the vernacular, it was summed up to some extent by *The Studio*, which, by the 1920s, was quite different to the radical new journal that had been launched in 1893 to champion the Arts and Crafts Movement. Neo-Georgian, neo-Regency, Tudor and Jacobean styles dominated its pages, but furniture manufacturers such as Waring and Gillow and Whiteley's, and pottery manufacturers such as Josiah Wedgwood, also responded to this taste, formulating a highly selective 'English' past. Without question, the deployment of these 'English' traditions proved to have widespread commercial appeal for large manufacturers and retailers, and as mass cultures they were particularly effective in representing 'Englishness'.

An idealized 'English' past was also conjured up by a continual engagement with traditional crafts that were intimately tied to the 'English' countryside. Gordon Russell's furniture company based at Broadway in the Cotswolds typified this, but significantly in the 1910s and '20s an interest in craft practices that were non-European invigorated the work of craftsmen and women such as the weaver Ethel Mairet, the printed textile designers Phyllis Barron and Dorothy Larcher, and the potters Bernard Leach, Michael Cardew, Reginald Wells and William Staite Murray. Japanese, Chinese, Korean and African ceramics and woven and printed textiles from Ceylon and Scandinavia were particularly inspirational, although 'English' vernacular traditions from before the Industrial Revolution were also researched and reinterpreted. Ironically, these new craft workshops, like their Arts and Crafts predecessors, were often located in idyllic rural settings, but depended upon an urban, metropolitan, sophisticated market for their economic survival.

Liberty & Co., drawing-room from *The Studio*, 1908.

Significantly, however, craft retained an experimental edge through important connections with the artistic avant-garde. In St Ives, Bernard Leach was part of a network that included the artists Ben and Winifred Nicholson and Barbara Hepworth; at Haslemere and Ditchling, Ethel Mairet knew and worked with the sculptor Eric Gill; and in Chelsea and Rotherhithe, William Staite Murray exhibited with the Seven and Five Society, of which Nicholson and Hepworth were members. These links helped to reinvigorate and redefine the crafts at a critical distance from nineteenth-century Arts and Crafts Movement theories and practices, which appeared increasingly outmoded and overly moralistic. The interest in design from around the globe overlapped, to a certain extent, with the enthusiasm for non-Western arts and crafts that was to be found within modernist circles. At this intersection of craft, design and the artistic avant-garde in Britain was the critic, potter and painter Roger Fry, who, via the First and Second Post-Impressionist exhibitions held at the Grafton Galleries in 1910 and 1912, introduced the British public to Post-Impressionism. A growing commitment to modernism was also articulated from 1915 onwards by organizations such as the Design and Industries Association (DIA) and the British Institute of Industrial Arts (BIIA), founded in 1920. Influenced by the perceived successes of the Deutsche Werkbund in stimulating better-quality industrial design and enhancing the relationships between manufacturers, consumers and the economy, the DIA was formed in 1915 in a context shaped by the First World War. A pamphlet from 1918 used the rhetoric of warfare and siege, to argue that

> Reconstruction . . . is our great need. British commerce has been
> entrenched for a century, but your competitors are everywhere

Bedroom suite in chestnut designed by Ambrose Heal and executed by Heal and Son, *The Studio*, 1906.

preparing to storm your entrenchments. Perhaps you think you are safe – we always do until the catastrophe, but, believe me, the years that are coming will be critical times. Technically, British goods are better than the foreigner's, but his chemists are experimenting day and night to catch up, and the British manufacturers' last line of defence will have to be design – living, progressing, and constantly developing. This may well prove to be the only way of keeping your lead.[5]

Britain and Post-war Reconstruction

At the Armistice in November 1918, some groups of people emerged from war impoverished by wartime inflation and the loss of principal breadwinners to the war effort. Others, such as young women and those who had previously been employed casually, benefited from changes in the job market – the former switching from domestic service to better paid work, and the latter benefiting from the huge demand for labour. Although there is disagreement over the impact of social reforms made during and immediately after the war, it has become more widely accepted that 'the absolute destitution which had haunted the poor of Edwardian Britain was banished'.[6]

Typically, the interwar years have been characterized as one long depression, and without a doubt the economic transformations that occurred during the period had the profoundest effect on design.[7] Following a brief boom in the years 1919–20, based on demand from industry for raw materials and from the consumer for long-imagined goods, industrial output fell by 10 per cent in 1921. There were several long-term problems for the economy, mainly caused by the decline of the British trade surplus, the emergence of the USA as the major player in the international economy and export problems for British staple industries – textiles, iron and steel, engineering, coal and shipbuilding.[8] Overall, Britain took a smaller section of

Designing Modern Britain

world export trade than it had before the war and its exporting industries were less competitive – in a market where trade did not recover to 1913 levels until 1925 and thereafter grew more slowly than before the war. But an underlying paradox was evident.

Characterized as a period of economic dislocation and instability, these years were also accompanied by significant economic growth.[9] To some extent, this growth – based on increased consumer expenditure and concentrated in certain industries that responded to this demand – blurred the divisions between the better-off end of the working class and the middle class. But divisions of other types were evident, between the old and the young, between those married and those single, and in geographical terms between the regions of Britain. These divisions became more acute because of the economic decline of certain regions, notably the north and west, which were dependent on staple industries. These areas, Clydeside, Tyneside, Lancashire and South Wales, were economically vulnerable, whereas Greater London and the Midlands were more prosperous, being based on the newer industries and services geared to the domestic consumer. By the end of the decade, however, an important structural shift away from export industries to the home market presaged well for the future.[10] The problem was that unemployment was experienced differentially, while the drift to the south-east of England for the expansion of newer industries – dependent on electricity rather than coal – accentuated regional inequalities that persisted through the 1930s and after the Second World War. Social class remained the sharpest of divides in 1920s Britain, 'one per cent of the population owned two thirds of the national wealth; and 0.1 per cent owned one third. Three quarters of the population owned less than £100.'[11] Nevertheless, there was a gradual and general increase in living standards over the course of the period, but poverty was still evident, resulting primarily from unemployment, supplemented by old age, chronic sickness, low pay and large families. A characteristic of the 1920s and '30s was increased consumption, and except for the poorest slum areas, it was uncommon to find children in ragged clothes and without shoes, as had occurred before 1914. Younger people in particular could dress relatively fashionably by making clothes or, when funds allowed, shopping at the cheaper multiples or at the Co-op. As several writers commented at the time, it was no longer apparent from looking at a person's clothes to which class they belonged, since companies such as Montague Burton Ltd and Prices, Tailors Ltd, dubbed 'the Rational Tailor', could, with the aid of new technologies, mass-produce a man's suit for 50 shillings.[12] By the mid-1930s Montague Burton had more than 400 branches and by 1931 Prices had 250.[13] These companies kept prices low by manufacturing their own lines and buying in only specialist ones. They charged low prices, advertised widely, priced garments very clearly and set up shop on main shopping streets so as to reach 'the man who cannot afford a suit'.[14]

New mass-production methods of this type highlighted the impact of America on British cultural life, as indicated by Priestley's mid-1930s diatribe against 'by-passes and housing estates and suburban villas and cocktail bars gleaming with chromium trim'.[15] The consumption of such goods shaped popular perceptions of America, but other types of goods referenced national identities, albeit in a contradictory fashion. A consciousness of 'British' goods coexisted with an awareness of those from the empire, Wales, Scotland and the regions.[16] For example, the Scotch Wool shop, first opened in Greenock in 1881, had expanded to 200 shops by 1910 and to 400 by 1939. These sold an image of Scotland – its knitwear and woollen products – as traditional, based on excellent quality and authentic materials. But paradoxically, new multiples such as Marks and Spencer, expanding from the mid-1920s, brought a corporate, modernist identity to Britain's high streets that was at odds with distinctive regional and national identities. The result 'was to erode the distinctiveness of provincial cultures, whether it be through the influence of mass advertising and commerce, or through the assertion of centrally-fixed standards and expertise'.[17] In parallel, and adding another layer of complexity, regional and national identities were articulated via consumption strategies that stressed a different set of qualities, including authenticity and quality. But, significantly, it was not just the countries of the British Isles or indeed the empire that were represented in this way, but also those of Ceylon, China, Korea, Japan and Scandinavia, contributing to a process of reinvigoration that particularly affected the crafts.

Experiencing Modernity: Everyday Life in the 1920s

Arriving in a variety of forms – magazines, advertising, cinema, dance, popular music, fashion – modernity changed the tempo of life. New transportation systems were crucial. The motor car remained expensive, but it was essential transport for the wealthy; consequently, there was increasing congestion on the roads, particularly in the centre of cities, such as the West End of London. Debate about road planning, bridge construction, garage building and the development of car parking preoccupied planners and authorities already concerned with the disorder of Britain's streets typified by the plethora of advertising hoardings. In most cities and towns across Britain, motorbuses, trams and trolley buses provided not just cheap travel for the masses, but mobile advertising displays. Ordering London's streets

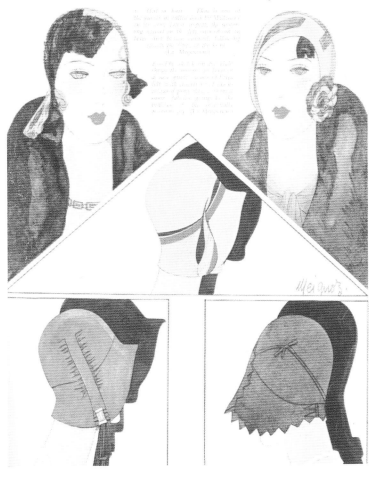

and transport became a priority in the 1920s, and in this respect Frank Pick, managing director of the Underground Electric Railways, played a central role. A founder member of the DIA and contributor to its campaigns against 'the untidy garage and the roadside hoarding', he initiated new station building for London's Underground system from the mid-1920s.[18] He employed the architect Charles Holden to design new stations in central London, such as Bond Street (1927), in addition to extensions to the Northern and Piccadilly lines towards the end of the 1920s. Pick and Holden were early enthusiasts for Continental modernism, and in 1930 they visited Germany to study new architecture. The apparent rationalism of

European architecture and design reinforced Pick's own thinking, itself informed by DIA ideas such as 'fitness for purpose'. Holden's new stations were recognizably 'standardized' in their simplified, modern appearance and technological innovation, albeit within a familiar monumental classical style. Pick aimed to bring visual coherence and unity to London's public transport system, employing artists and designers to produce posters, rolling stock, textiles and typography as well as stations. But it was not until the mid-1930s that this new corporate identity took on a recognizably 'international' modern style.

Bond Street underground station, designed by Charles Holden, 1927.

Increasingly visible in the underground and on the buses were women going about their everyday lives. Magazines such as *Britannia and Eve*, *Woman and Home* and *The Lady*, and fashion monthlies such as *Vogue*, proved to be effective in responding to the 'new woman'. An array of goods for the female body and for the home were illustrated, discussed and promoted with the ultimate aim of keeping the modern women up-to-date. Advertising was particularly effective. An ad of 1927 promoting new electrical products showed a young and fashionable housewife using a vacuum cleaner and being advised to 'let electricity do your drudgery. Don't do it yourself.'[19] Other advertisements promoted modern materials such as Celanese, an artificial spun silk that offered the advantages of warmth, lightness and wearability.[20] Modernity was a constant theme. The modern body was to be lean and muscular, as seen in a *Britannia and Eve* advertisement for Kestos Lingerie. In 1929 the magazine promoted exercises derived from a 'primitive' dance for 'the only golden road to a good figure';[21] in the same issue it appealed to a wider market with an advertisement for Helena Rubenstein, which explained that with the aid of new laboratories and salons, 'every woman – no matter how little she is prepared to spend – shall have the advantage of my worldwide experience and research'.[22] The self-help features in these fashion pages were part of a culture of manipulation and transformation that located the female body as artificial and malleable in pursuit of an ideal. The emphasis was on the *science* of cosmetics and the *profession* of beauty care. Managing your appearance was comparable with managing the home – both required new skills that could be learned; *Woman and Home* in December 1926 had advice for 'Keeping Up

Designing Modern Britain

Bathroom with portable electric fire and electric towel rail, Reyrolle publicity brochure, Hebburn-on-Tyne, 1927.

Reyrolle publicity brochure, Hebburn-on-Tyne, 1927. 'Let Electricity do your drudgery. Don't do it yourself!'

Appearances', including new and well-tried beauty techniques – professional massages, shingle caps (for keeping the new style in place), plus advice for combining earrings and Eton crops. To be *feminine* depended upon being *modern*. Hair, in particular, was a sign of modernity and the magazines gave advice on various short styles; particularly fashionable in 1926 was the shingle, cut very short at the back with waves over the ears.

'Englishness and Identity'

Wearing sharply cut tailor-mades, the women depicted in these magazines were far removed from those only 15 years earlier, when fashionable women were voluptuous, Junoesque figures clad in rich and extravagant robes. In these 1920s magazines, the ideal was boyish and slim with little in the way of hips or bust. The style emerged in the late 1900s and early 1910s from the Parisian couture house of Paul Poiret, but via eighteenth-century *directoire* lines, Russian art and culture, and Indian textiles. By the 1920s Poiret's prototype elongated outline had become increasingly androgynous and exaggerated in the hands of Patou, Chanel and Lanvin. The ideal was best observed in high-fashion magazines such as *Vogue*, first published in England in 1916, and *Harper's Bazaar*, first published in 1929.

Woman using curling tongs, Reyrolle publicity brochure, Hebburn-on-Tyne, 1927.

The very wealthy bought their gowns from the Paris couture houses of Poiret, Lanvin, Callot Soeurs and Patou, but if Paris was out of reach (materially and geographically – as it often was), it was possible to acquire couture clothes from specialist sections of department stores such as the French Salon at Fenwick in Newcastle and from one-off boutiques, either in the regions or in London. These fashionable young women, whose slender bodies came to characterize a decade, benefited from the changes wrought by war and the vote; although they were hardly representative of women generally, they nevertheless epitomized a sense of freedom and derring-do. Women such as Miss Smith, the daughter of John Smith, owner of Smith's Foundry in Barnard Castle, were examples. After her mother died in 1925, she travelled extensively in Italy and France, never marrying, but, with a reputation for dressing outrageously and eccentrically, she frequented the fashionable races at Longchamps and Auteuil in the mid-1920s.[23] Slightly older but still part of this new generation of modern women were the local aristocrats Lady Danesfort and Linda Rhodes-Moorhouse. Danesfort's collection of costume donated to the Bowes Museum included gowns by Paquin and Reville (Paris and London). Dressed in Lucile for her wedding in 1912, Linda Rhodes-Moorhouse was from the Morritt family that owned the Rokeby Park estate, County Durham. Typical of her generation of women, her husband died in

Advertisement for Kestos Brassiere and Girdle, *Britannia & Eve*, November 1933.

Designing Modern Britain

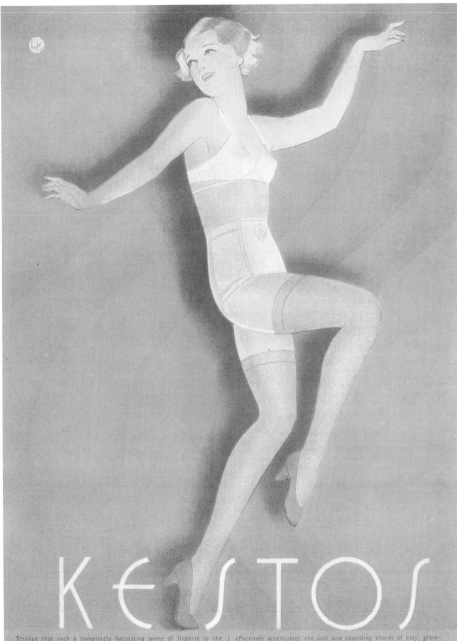

KESTOS

Strange that such a completely fascinating piece of lingerie as the Kestos Brassière should possess the ability to endow the feminine form with so much loveliness, yet never has a woman found a more reliable aid in her quest for beauty. Worn next to the skin, the Kestos Brassière skilfully controls over-emphatic curves and most effectively accentuates the soft and appealing charm of easy, graceful contours. Wholly captivating, too, are the girdles fashioned by Kestos, and so completely successful are their powers of persuasion that unlovely curves and distressing fullness are almost magically transformed into rhythmic lines.

Kestos Brassière, 3/11 to 17/11. Kestos Girdles at various prices. Sold everywhere.

Catalogue on request. KESTOS LTD., Maddox House, Regent Street, London, W.1. (Wholesale only.) Also at Paris and Brussels.

the First World War and her son was killed in the Second. Although these women would have spent most of their time in London, as members of the upper class in north-east England, they were admired and observed locally, in addition to attending important civic functions and regional events. Attending one such event, the Mayor's Ball in Darlington, was Marie Thirkell; born at the start of the twentieth century, she was married to the blacksmith in the nearby village of Middleton Tyas. Reputedly given five sovereigns by her husband, she bought a ballgown at Binns department store in Darlington, in which she looked 'like a chandelier going round'.[24]

A characteristic of fashionable styles in the 1920s was widespread dissemination. The sleek, attenuated style was hugely popular and women from all social classes dressed in the cylindrical fashion. A series of photographs of South Park in Darlington shows several young women at leisure, either walking or on the putting green, wearing the fashionable styles of shortened skirts, casual loosely fitted jackets and T-bar shoes. Even young women workers at Pease's Woollen Mill, dressed in their finest for a works' dance, wore sheath-like dresses, shortened skirts and casual cardigans. With bobbed and shingled hair and silk stockings, they were not that far removed from their couture-clad counterparts. Most illustrations of these styles in *Vogue* and other women's magazines depict an idealized female form – slim, flat-chested and tall – but as these photographs reveal, naturally rounded bodies had to be 'shaped' into the angular, tubular styles with the aid of flattening corsets. Although constrained by social class more effectively than corsets, these young women shared in a sense of new identity

Designing Modern Britain

that linked them to metropolitan London and beyond. Between 1914 and 1929 fashion 'imagined' modernity, and although the style had its origins in the past, the 1920s look owed little to this.

Re-imagining England: Collecting, Craft Practice and Omega

Between 1910 and 1929 manufacturers and designers of both industrial and handmade ceramics began to explore traditions of design that were either pre-industrial or from non-Western cultures. This was in response to, but it also stimulated, the development of new scholarship and patterns of collecting. Early Chinese and early English ceramics began to be admired and studied within a context of questioning regarding design standards and taste. An important contributor to this was Roger Fry, who, as editor of the *Burlington Magazine* between 1910 and 1919, published numerous articles on ceramics from a range of different cultures – Peruvian, Chinese, Korean, Japanese and African – but Fry was particularly influential as the founder of the Omega Workshops. Omega was directly inspired by modernist ideas and aesthetics, and inevitably this provided another conduit for the wider dissemination of information about the art and artefacts of non-European cultures. At the same time, studio ceramicists such as William Staite Murray studied ceramics from China and Korea as a way of rethinking the relationship between art and craft. Marking a clear break with the nineteenth-century Arts and Crafts Movement, Murray and a number of others developed an aesthetic borne of abstraction based on stunning glaze effects and simplified forms.

The taste for new categories of ceramics was both a reaction and a contribution to the growing appreciation of modernist art and design and a reassessment of 'English' design. Pre-industrial ceramics (Chinese and English), studio ceramics and modernist art and design were perceived to share certain qualities, including formal abstraction, sparse (sometimes rough) forms, simplified use of colour, minimal decoration (the ceramics were often simply incised or carved with glazing that appeared accidental, often including streaks, drops, pools, cracks and crazing) and a growing concern for the spiritual dimension of art and design. Fry recognized in these ceramics qualities that were absent from European art and design:

> we are more disillusioned, more tired of our own tradition, which seems to have landed us at length in a too frequent representation of the obvious or the sensational. To us the art of the East presents the hope of discovering a more spiritual, more expressive idea of design.[25]

The *Burlington Magazine* was the focus for a network of scholars, collectors and theorists who contributed to this shift in taste. It included Fry, an art theorist and occasional potter; the writers and curators Robert Lockhart Hobson (Keeper of Ceramics and Ethnography at the British Museum) and Bernard Rackham (Keeper of Ceramics at the Victoria and Albert Museum); and the collectors George Eumorfopoulos and William Alexander, whose taste for early Chinese ceramics (Han, Tang and Song as opposed to Ming or Japanese ceramics) began to develop at the end of the nineteenth century and the beginning of the twentieth. Although an appreciation of Japanese and Chinese ceramics was evident in the last quarter of the nineteenth century, it was mainly eighteenth-century porcelains that were admired.[26] Interest in early Chinese stonewares, porcelains and celadons, particularly those of the Tang (AD 907–60) and Song (960–1279) dynasties, was stimulated by excavation from tombs during railway building in China in the early years of the twentieth century. Between 1909 and 1920 early Chinese ceramics were comprehensively described and illustrated in the pages of the *Burlington Magazine* by Hobson, Rackham and Fry, but a serious scholarly interest in pre-industrial ceramics had existed in the magazine since 1903. There was also substantial interest in English pottery, but generally that made before the technological changes of the eighteenth century; early English stonewares and slipwares, and early Staffordshire pottery. This academic interest was complemented by the activities of a group of young studio potters, including William Staite Murray, Reginald Wells and Nell and Charles Vyse, as well as the pottery manufacturers Bernard Moore, who experimented with oriental glaze effects and employed John Adams and Dora Billington; Doulton (in Burslem), which employed Phoebe Stabler and Charles Vyse; and Adams and Stabler (via the Poole Pottery, the manufacturer of Fry's Omega pottery).

Studio pottery, which had begun to develop in the 1910s, was significantly different from the ceramics of Arts and Crafts Movement potters such as William De Morgan. It was fundamentally abstract, and although it shared an interest in non-Western cultures, it was more concerned with form and glazes than pattern and decoration. For the studio potter William Staite Murray, 'the point of the pot' was that it combined creative processes with the opportunity for aesthetic, imaginative and spiritual expression, thus signalling a shift in emphasis from the socially oriented Arts and Crafts Movement.[27] Murray's practice as studio potter, which began in 1909, was suffused with oriental ways of working and thinking. As a student at the Camberwell School of Arts and Crafts around 1909–10, he had been inspired by the early Chinese pottery that was illustrated and discussed in the pages of the *Burlington Magazine* and seen at the

Bowl with light brown body and grey-white glaze designed and made by William Staite-Murray, 1920s.

A thirteenth-century bottle illustrated in 'The Art of Pottery in England' in *The Burlington Magazine*, XXIV (October 1913–March 1914).

important exhibition *Early Chinese Pottery and Porcelain* held at the Burlington Fine Arts Club in 1910. His fellow potter Bernard Moore worked for years perfecting the Chün glazes typical of Song ceramics, while Bernard Leach had first-hand knowledge of Japan, having spent several long periods there between 1909 and 1919, and of China, living from 1914 to 1916 in the Northern Provinces, where he saw Tang and Song ceramics. Murray, Wells and Charles and Nell Vyse lived near the collector George Eumorfopoulos and gained first-hand experience of early Chinese ceramics from his pioneering private collection. Eumorfopoulos, who built a two-storey museum to house his collection at the back of his house in Chelsea, had first seen the few specimens of tomb wares to reach the West at the Burlington Exhibition in 1910; to him and others, these functioned as a 'corrective to Western one-sidedness'.[28]

Writing on ceramics in the *Burlington Magazine* paralleled the shifting tastes of collectors. A series on 'Early Staffordshire Wares' begun in 1903 by Hobson focused on ceramics prior to the industrialization of the pottery industry. In the first article he examined Staffordshire slipwares, and in his depiction of England as primitive, 'wooded, wild and picturesque', he evoked an image of ceramics production as organic and integral to everyday life, focused on the individual potter. This representation of the solitary potter was to inspire subsequent studio potters such as Murray and Leach.[29] Hobson identified the qualities of these Staffordshire slipwares that were especially appealing: 'there is something genuinely fascinating in their naive simplicity and their entire lack of all that is artificial or extraneous . . . These early pots are like the potters who made them and their friends who used them, English to the back-bone.'[30] This emphasis on their 'Englishness', and on the robustness and simplicity of design, was particularly revealing. Timelessness and authenticity were implicit in this definition of 'Englishness', in marked contrast to the vagaries of fashion and stylistic invention typified by ceramics after the Industrial Revolution. The *bête noir* of writers such as Hobson were industrial ceramics produced in Stoke-on-Trent, in particular by Josiah Wedgwood, who in the 'year 1759 . . . parted company with old Staffordshire traditions'.[31] Hobson's view that Wedgwood was synonymous with the loss of traditional pottery-making values was reiterated by Fry in the *Athenaeum* in 1905:

Earthenware mug painted with underglaze blue and purple lustre, designed by Louise Powell for Josiah Wedgwood & Sons Ltd, 1910–1920. These wares marked an attempt to synthesize Arts and Crafts ideas and practices with modern industrial production methods.

> Wedgwood pottery contributed to the final destruction of the art, as an art, in England, because it set a standard of mechanical perfection which to this day prevents the trade from accepting any work in which the natural beauties of the material are not carefully obliterated by mechanical means.[32]

An interest in ceramics from other cultures developed in tandem with this concern for traditional pre-industrial English pottery, and there were clear aesthetic parallels: lustre wares from Persia and Egypt, maiolicas from Italy, 'Hispano-Moresque' pottery from southern Spain, and Peruvian pottery all shared a certain naivety, simplicity and robustness in terms of form, colour and decoration.[33] Hobson's discussions of Chinese and Japanese ceramics in the *Burlington Magazine* appeared alongside these. He was entranced by the simplicity and beauty of early Chinese wares, noting the variety of glazes such as those 'that were thick, like congealed fat', and the stunning glaze effects: accidental splashes of iridescent lilacs and rusty red, speckle and streaked glazes and crackle effects such as 'crab's claw crackle' and 'fish roe crackle'.[34] The sheer detail and excellent quality of illustrations provided would-be studio potters with an immense amount of information.

Other writers contributed to this, such as Bernard Rackham, who played an important role in putting together the Victoria and Albert Museum's ceramics collection during this period; he acquired Chinese

Designing Modern Britain

ceramics as well as work by Reginald Wells and William Staite Murray, the latter received as gifts from Eumorfopoulos. Like Fry, Rackham believed that ceramics were 'first and foremost, a plastic art' rather than a representational one. Fry's inclusion of ceramics by Matisse, Vlaminck and Derain at the First Post-Impressionist exhibition in 1910 had enabled him to make 'a degree of abstraction in painting more comprehensible by putting it alongside work in a medium where abstraction was already accepted'.[35] Later, when contemplating a Song bowl, Fry noted its external and internal contours, its density and resistance, colour and glaze, before finally observing 'the pot is the expression of an idea in the artist's mind'.[36] His critical essays on English and oriental ceramics in the *Burlington Magazine* were clearly building blocks in the development and formulation of his aesthetic position, which brought together art and design.[37] Focusing on 'English' qualities in ceramics, Fry's essay 'The Art of Pottery in England' of 1913–14 discussed the relationships between great art or craft and the society in which it was produced. Attempting to understand the superiority (in his mind) of early forms of English pottery over later forms, he believed that creative effort in the craftsman and fine appreciation in the public were unable to 'arise freely in a wholly degraded and brutal society'.[38] Significantly, Fry drew comparison between English thirteenth-century pottery and Tang wares:

> It has not quite the subtle perfection of rhythm in the contour, and the decoration is rather rougher and less carefully meditated. But to be able to compare it at all with some of the greatest ceramics in existence is to show how exquisite a sense of structural design the English Craftsman possessed.[39]

Probably crucial to Fry's position in relation to ceramics was this notion that 'it is expressive of what we instinctively recognize as a right state of mind'. This ideal of a craftsman completely at one with his chosen craft, combined with the interest in the medieval, reveals the influence of Arts and Crafts thinking, but Fry's concern for particular formal elements signalled a clear difference. Generally, Fry thought abstractly about design, and it is revealing that when he wrote about Chinese ceramics he emphasized the sculptural rather than the pictorial, the useful rather than the decorative, and individual, spiritual expression in relation to ceramic bodies and glazes rather than technical perfection. Like several of his contemporaries, he leaned towards the earlier periods of Chinese ceramics, such as Tang and Song, as opposed to Ming, citing the 'blunted edges and the absence of mechanical evenness in the surface of planes'.[40] Such aesthetic preferences informed his activities at Omega Workshops; in a prospectus of 1915 he wrote that pottery was 'essentially a form of sculpture', and its surface should 'express directly the artist's sensibility'.[41]

At Omega, Fry revealed a commitment to practice as well as theory. From the beginning, in 1913, he made pottery by hand-throwing on the wheel for hand-painting by Omega artists.[42] He had taken lessons from a traditional flowerpot maker in Mitcham, south London, as well as attending Camberwell School of Art in the autumn of 1913, where he was taught by Richard Lunn. The Head of Camberwell School of Art was W. B. Dalton, whose serious interest in ceramics led him to experiment with Chinese stoneware and to establish the school as one of the most influential for the study and practice of ceramics (Murray and Wells had attended before Fry, and Charles Vyse taught there).[43] With practical training in hand, in 1914 Fry set about extending Omega's product range from hand-painted bowls, vases and jugs to tea, coffee and dinner sets by establishing a working relationship with the Poole Pottery in Dorset. At Poole, Fry had moulds made to his design, enabling Omega ceramics to be made in larger numbers. For Fry, the machine and mass-production meant that 'the nervous tremor of the creator disappears', but at Poole he found the ideal company ethos and experimental context to enable such problems to be overcome. For example, all moulds were initially made by hand by a craftsman, and to Fry this ameliorated, to some extent, the negative effects of mass-production.[44] The Poole Pottery had been established as a producer of architectural terracotta in the mid-nineteenth century, but at the time that Fry was involved with the company it had just developed a new range of tablewares using an off-white stoneware dipped in a creamy-white slip and covered with a clear glaze, which, when fired, fused to give a soft blotted effect. The company strove for a handmade quality, and the use of stoneware, small-scale pro-

Tea and coffee set, earthenware with white tin glaze, designed by Roger Fry, 1916–18, at the Omega Workshops.

Designing Modern Britain

duction and hand-painting contributed to this. Judging by its appearance, Omega pottery from 1914 onwards obviously benefited from this experimental context. The white, blue and black glazes and simple shapes have a definite 'Chinese' pre-industrial feel in terms of glaze effects and form, and their simplicity is quite at odds with mass-produced pottery of that date.

Omega was essentially contradictory, and Fry's position summed this up. Largely at odds with mass-production, Omega goods also flouted most of the procedures and processes associated with good craftsmanship.[45] While Omega's commitment to domestic design and applied art linked it to the reformist strategies of Morris and Ruskin, visually it looked to French modernist art. Indeed, through Omega, the well-off could buy textiles, furniture, murals, screens and interiors – designed by the young avant-garde of British art including Duncan Grant and Vanessa Bell – whose formal origins lay in the paintings of Manet, Matisse and Picasso, rather than Rossetti or Burne-Jones. Colour was vivid; patterns were bold and decorative; and forms were abstract or loosely figurative, and relatively simple. Fry's preferences for vernacular or austere classically shaped forms were evident in his furniture designs at Omega, which also referenced late Arts and Crafts design. Admiring the 'expressive' qualities of particular periods of English ceramics, rather than intricate craft skill, the rough finish of Omega products is explicable. Although short-lived (it closed in 1919), the Omega Workshops were part of an important reorientation of design in the period just before and after the First World War. Craft practices were reinvigorated, but alongside this there was a growing awareness of European and non-European art and design, reinforcing emerging modernist aesthetics. Fry remained an important figure in British artistic and cultural life during the 1920s, producing a number of influential essays and contributing to various government initiatives to improve design standards.

There were numerous links between the various strands of craft activity, its practitioners, collectors and apologists. Arts and Crafts teaching and practice was perpetuated in the art schools – Camberwell School of Art being a case in point; traditional 'English' and non-European crafts were collected, researched and exhibited – Chinese ceramics were a good example; and new initiatives in the crafts emerged particularly in ceramics and textiles typified by the work of Murray, Leach and Mairet. Alongside and interlinking with this were emerging modernist discourses that synthesized with Fry's notion of authentic 'Englishness'. Omega was clearly representative of a particular strand, but less orthodox links developed in this climate of experimentation. The rediscovery of specific English traditions 'gave craft and, by association, the handmade surface, a nationalistic dimension', which was also paralleled in modernism.[46] In examining the

relationships between modernism and particular native English craft tradi-
tions, there is a case for rethinking the relationships between modernism,
'Englishness' and craft in Britain in the 1910s and '20s. The work of William
Staite Murray enables a useful exploration of the nuances of modernism in
design along a trajectory that includes the Omega Workshops, the Arts and
Crafts Movement, English vernacular forms and non-European ones.
Murray was linked to Omega through the artist Cuthbert Hamilton, and he
exhibited with modernist artists, including Ben and Winifred Nicholson.
Hamilton and Fry worked together at the Yeoman pottery from 1915, producing pieces that were formally related to contemporary work by Vanessa
Bell and Duncan Grant, but it was Kandinsky's work in particular that
Murray admired.[47] Like several of his contemporaries, Murray's passion for
early forms of Chinese and Korean ceramics led him to try to create ceramic bodies, glazes and forms that captured some of their qualities, but Fry's
writings on early English ceramics, Chinese ceramics and modernist art
also offered a theoretical rationale for Murray's distinctive approach. Like
Fry, Murray was committed to the integration of art and design:

> paintings and sculpture as well as pottery, suffer through being con-
> sidered as independent units, instead of part of an organised
> decorative whole . . . the tendency of modern art exhibitions is to
> show paintings, sculpture and pottery together, and not separately.[48]

To Murray, pottery forms were to be contemplated as abstract art, rather
than utilitarian objects, and thus by exhibiting them alongside the work of
painters and sculptors he believed the public would learn to think concep-
tually about design, painting and sculpture.[49] As a member of the Seven
and Five Society from 1927 and a frequent contributor to exhibitions with
artists and sculptors such as Ben and Winifred Nicholson, Paul Nash and
Jacob Epstein, Murray believed that 'pottery may be considered the connecting link between Sculpture and Painting, for it incorporates both'.[50]
Such ideas demonstrate that modernist thinking had permeated British
visual culture considerably earlier than the 1930s.

Reworking the Eighteenth Century: Ceramics and Furniture

A concern for specific periods of history and particular representations of
the 'English' past also characterized the activities of a number of manufacturers and retailers in the first decades of the twentieth century, notably the
ceramics manufacturer Josiah Wedgwood and Sons Ltd and the furniture
producers Waring and Gillow and Whiteley's. In part this was in response

to changes in the market, but it also connected with the interest in English traditions that came to the fore within the context of the Arts and Crafts Movement and subsequently. As Arts and Crafts architects and designers such as Edwin Lutyens turned to classicism in the 1910s and '20s, a discernible shift towards eighteenth-century styles of design took place more widely. More than 150 years in existence, Josiah Wedgwood celebrated and re-examined its eighteenth-century origins as it attempted to establish new markets and to consolidate existing ones prior to and just after the First World War. Representing less than 1 per cent of British manufacturing, the pottery industry had reached maturity by the end of the nineteenth century. Highly dependent on the export market, which absorbed one-third of output, after 1914 the home market gained in significance, and under the art director John Goodwin, a programme of rationalization and an overhaul of the product range took place.[51] Elaborate Victorian wares were withdrawn, and new shapes and patterns were developed that echoed the company's original eighteenth-century designs. In 1906 the company had exhibited a number of revived eighteenth-century designs that had been suggested by the Arts and Crafts designer Louise Powell, who, along with her husband Alfred, had been hired on a freelance basis. She had come across some of the original pattern books during one of her visits to the Etruria factory in Stoke-on-Trent. These designs, modified to suit contemporary tastes, were in a traditional manner involving a large amount of free-hand painting. The revived eighteenth-century patterns proved a commercial success thanks to the growing appreciation of Georgian and Regency design in the first two decades of the twentieth century. The patterns, including 'Vine', 'Oak Leaf' and 'Crimped Ribbon and Wreath', were sold through mainstream stockists such as James Powell of Whitefriars and Harrod's, although later in the 1930s more progressive modern stores such as Dunbar Hay stocked them too. This was part of a strategy to produce attractive tableware for an expanding middle-class market, and the eighteenth century was regarded as a benchmark for good taste after the apparent excesses of the Victorian period. Relatively simple lines, uncluttered surfaces and good proportions were typical of ceramics, furniture and architecture from this period, and they were widely admired and copied. The interwar period saw an ongoing interest in Georgian architecture and design leading to the formation of The Georgian Group in 1937, which campaigned to protect and preserve buildings, monuments and gardens in the face of unremitting destruction. Frequently derided by those modernists who were inspired by the rather didactic approach of Continental exponents, retrospectively it is clear that with its emphasis on geometry, form and simplicity, Georgian architecture and design provided a design language not that dissimilar in some respects to Continental modernism, but with broader appeal to designers, manufacturers and consumers as a result of its connection with 'English' traditions.

Looking back to the eighteenth century was certainly effective promotion and marketing for a company that had secured its premier position in English ceramic manufacture in that period, and could justifiably claim a lineage of excellence in design and manufacturing. Equally true to its eighteenth-century precedent, Wedgwood employed artists and designers to help reorient it. The French designers Marcel Goupy and Paul Follot had been employed towards the end of the nineteenth century, and in 1903 Alfred and Louise Powell were employed to synthesize Arts and Crafts principles and progressive design practice. The Powells took their cue from both medieval and eighteenth-century prototypes, as had many of their Arts and Crafts counterparts. Most of their designs were free-hand-painted jugs, bowls, plaques, vases and small covered pots. They were decorated in bold, stylized patterns derived from nature, but with a loose graphic quality. In a manner similar to art and studio potters, they too looked further afield than just the 'English' past, exploring early Italian and Renaissance pottery, particularly in terms of colour, form and glaze. At Wedgwood, their use of lustres was inspired by Italian tin-glazed maiolica and Persian ceramics. Shapes were vernacular, and patterns suggested either medieval, heraldic forms or they were based on small repeats of flowers and leaves handled in an eighteenth-century manner. By the mid-1920s fifty of their designs had gone into production, and the impact of their ideas was felt across the company as the range moved towards a simpler, hand-finished aesthetic that helped to re-establish the company's credentials for good design. Writing in *The Studio* in 1929, Alfred Powell argued that it was 'really no easy matter at the present time to find a piece of modern industrial pottery that could be called by a properly constituted judge "beautifully painted"'.[52] Grace Barnsley, daughter of the Arts and Crafts designer Sidney Barnsley, and Louise Powell's sister Thérèse Lessore both worked at Wedgwood as designers; the work of the latter was unusual because of its close formal connections to the modernist-inspired Bloomsbury artists Duncan Grant, Vanessa Bell and Roger Fry, and to Omega. An earthenware tea set in grey and pink lustre done in the late 1910s that depicted a bather disrobing owed a great deal to Grant and Bell with its loosely handled paint and formal abstraction.

Josiah Wedgwood's enthusiasm for the firm's eighteenth-century origins was evident in a book the company published in 1920, entitled *Arts Etruriae Renascunter: A Record of the Historical Old Pottery Works of Messrs Josiah Wedgwood & Sons Ltd.* Inside, a subtitle noted that the book was a record of the old works 'as they exist today, forming an unique example of an eighteenth century English factory'.[53] The narrative and the accompany-

Designing Modern Britain

ing illustrations, evoking the company's history and achievements, were the work of staff from the art department, Harry Barnard and James Hodgkiss. As an account of what remained of the old Etruria factory in 1920, it self-consciously referenced tradition and authenticity to reinforce Josiah Wedgwood's significance as a design and manufacturing pioneer:

> The atmosphere and tradition of the late eighteenth century, carefully and faithfully preserved through five generations to the present one of both master and man, is the real factor that enables the present-day production to maintain that high standard which the untiring energy and application of the original 'Master Potter' set out to accomplish.[54]

The wider context for this book was unquestionably post-war reconstruction and the imperative to demonstrate the company's historical significance, but also its future viability. At this modern Etruria:

> We are back again in the twentieth century, and Etruria will still be found to hold its own in progress which the present day demands from it, without losing the particular style that has proved at all times the valuable legacy of the illustrious founder.[55]

Aspects of pre-nineteenth-century English design were researched assiduously from the early twentieth century. Numerous books aimed at different readers were published, and as at Wedgwood, 'the new evaluation of eighteenth-century furniture was closely linked with new designs in these styles'.[56] Writers such as Percy Macquoid, W. A. Symonds and Frederick Litchfield began producing histories of furniture around the turn of the twentieth century, contributing to this, as did the publication *Country Life*, established by Edward Hudson in 1897.[57] *Country Life* was a hybrid magazine combining news on farming, property, dogs and hunting, plus selected product advertising (for example, Burberry outdoor wear) for the aristocracy and upper classes, with information about traditional rural life. The issue of 2 May 1914 was typical. With a front-cover photograph of Lady Evelyn King, inside it had features on May blossom and English orchards, highland terriers, 'Ten Famous Old Inns', and the lost craft of laying a hawthorn hedge.[58] In this particular issue, the architectural focus was French châteaux, but more usual was the December issue of the same year, featuring Arundel Castle due to its fine Georgian furniture.[59] From the outset, *Country Life* included extensive coverage of the country homes and estates of Britain's ruling elites and newly wealthy. Typically, a feature would examine the house, its history and its contents in some detail, and the text would be accompanied with high-quality black-and-white photographs that 'influenced the arrangement of rooms, encouraging for example the removal of groups of miscellaneous furniture'.[60] The stripping away of largely Victorian furniture and fittings and a re-emphasis on the original interior and layout was apparent in two photographs of Nostell Priory in Yorkshire. Designed by James Paine and Robert Adam between 1735 and 1780 with furniture by Thomas Chippendale, it was photographed for *Country Life* in 1907 and 1914. In the latter, 'the Victorian clutter had been removed and the purity of the Adam design re-emphasised'.[61] This concern for authenticity was reinforced by growing scholarship and the re-evaluation of particular styles – Georgian being a good example – and *Country Life* contributed to the growing taste for certain historical styles – English, especially, and certainly before Queen Victoria's reign. The enthusiasm for old houses was roughly divided between earlier 'English' styles such as medieval, Tudor and Jacobean, and classical styles from the late seventeenth to the early nineteenth centuries.[62] Equally, 'the taste for old furniture and old materials encouraged the taste for old houses, for restoration and period fittings, and vice versa'.[63]

From 1897 *Country Life* photographed and discussed a succession of restored country houses alongside a few new ones built in the preferred styles. In 'Country Homes: Gardens Old and New' (a regular feature), a number of important country houses and their interiors were discussed in detail, and although the initial focus was English examples, Scottish and Irish houses had begun to appear by the 1910s. Eighteenth-century Stowe was featured in January 1914; the Drum in Midlothian, described as 'the masterpiece of William Adam', appeared in October 1915; and Castletown in County Kilkenny of *c.* 1770 was featured in September 1918.[64] Alongside these were articles on furniture by Percy Macquoid. Writing over several months in 1919, he discussed 'English Tables from 1600–1800', including early Jacobean and Elizabethan, walnut from the Charles II period, and early Georgian and later eighteenth-century tables.[65] Parallel to this was a surprising interest in modernity. The efficiency of electric cookers – both commercially and domestically – was commented upon, particularly in relation to 'the present dearth of domestic labour'. An article in the same year, 1920, discussed the development of Dormanstown in Middlesbrough, a scheme for 3,000 houses developed by the steel company Dorman Long for its workers. Using new concrete construction methods, it was designed in a neo-Georgian style by the architects, Adshead, Ramsey and Abercrombie.[66] The neo-Georgian became particularly admired in the 1920s for its restraint and simplicity, evident in the new Garden City at Welwyn, north of London, but also in the building of the new country

Neo-Georgian houses at Dormanstown, Middlesbrough, 1920. Designed by Adshead, Ramsey & Abercrombie for the steel manufacturer, Dorman Long.

Interior of one of the Neo-Georgian houses at Dormanstown. Note the large open sitting room with reinforced concrete lintels.

houses that were featured in *Country Life*. Responding, no doubt, to the diminution of the aristocracy following the First World War and the expansion of the upper-middle class, by the 1920s the magazine had a new feature: 'Lesser Country Houses of Today'. Hownhall, Ross-shire, by Adshead, Ramsey and Abercrombie was a typical example. The house was essentially a 1920s reworking of late Georgian, and its architects displayed 'a keen discernment of the merits enshrined in late Georgian houses . . . These houses were pleasant to look upon, quiet and refined, and free of all freakishness and make believe.'[67]

The contribution of *Country Life* to notions of good taste in the 1910s and '20s was significant, and it exposed the increasing preoccupation with simplicity, which by the end of the 1920s overlapped with modernism. The backdrop to this was the disavowal of Victorian taste and the proliferation of information about the Modern Movement in journals such as *Architectural Review* and *The Studio*. Between 1914 and 1925 few examples of Victorian architecture and design were featured in the pages of *Country Life*,

Front elevation of 'Hownhall', Ross-shire, a 'lesser country house of today' according to *Country Life*, 1920. Designed by Adshead, Ramsey & Abercrombie.

Interior of 'Hownhall', Ross-shire. The neo-Georgian provided a benchmark for 'good' design for working-class and upper-class housing. Discernible characteristics were simplicity, refinement and a lack of preence.

Designing Modern Britain

whereas neo-Georgian was evident in photographs of workers' housing, Garden Cities and suburbs, and new country houses. These identified a new simplicity, which moved toward modernism via a simplified classicism and Scandinavian prototypes.

Good Taste, Retailing and the Domestic Interior

The question of taste became enmeshed with that of good design in the 1910s and '20s, as campaigners sought to improve the quality of manufactured goods and raise public consciousness on the issue. The pages of *Country Life* demonstrate that Chinese ceramics and Persian pottery, Georgian furniture and architecture were emblematic of high cultural values and provided a benchmark of good taste. But at the commercial end of the market, it was retailers and manufacturers who articulated public taste by leading and responding to consumer demand.

Advertising new premises with interiors of 'the rich Georgian type' on the recently developed Oxford Street, the furniture retailer and manufacturer Waring and Gillow also sought to stress its modernity. Referring to its factories in Hammersmith, Liverpool, Lancaster and Paris, it advertised 'modern labour-saving machinery and every up-to-date appliance for perfect and expeditious work'. Affirming its commitment to the twentieth century with 'its vivid modernity', the firm also traded on authenticity and continuity, combining 'modern methods, and the resources of modern machinery, with what one may venture to call a splendid ancestry'.[68] As it advertised the recently constructed 'New Galleries', it set out to convince customers (in a manner reminiscent of Josiah Wedgwood and Sons Ltd) that 'the old traditions are still the ruling principle'. Illustrations depicted room settings resplendent with

> delightful reproductions of old models, faithful copies of historic rooms . . . a hundred-and-fifty beautiful specimen rooms in every recognisable style, each one an exquisite ensemble of colour and an example of perfect taste, a range of completely furnished houses at inclusive prices, and through all and over all there run the predominating notes of refinement and economy.[69]

The Lancaster firm of Gillow, like Wedgwood, was an eighteenth-century company (merged in 1903 with S. J. Waring, Cabinet Makers) with celebrated connections to furniture makers such as Thomas Chippendale. Receiving commissions for large-scale public and commercial projects, including the ocean liner *Lusitania* and the Carlton Hotel in Pall Mall

The Georgian Library, part of the Waring & Gillow display at the British Empire Exhibition at Wembley, 1924.

The 1924 Bedroom, part of the Waring & Gillow display at Wembley.

(1899), the firm also produced a number of publications to cater for the more modest needs of the furniture-buying public. In a brochure from around 1910 for those desiring a 'refined home', it included a 'series of short articles illustrating the different styles of furniture used in England in past ages and the adaptation of these styles to our present day domestic requirements'.[70] Surveying 'the Oak Age', 'Elizabethan', 'Jacobean', 'William and

Designing Modern Britain

Mary' and various eighteenth-century styles of furniture, it provided a guide for the discerning customer in identifying and selecting different 'English' styles. 'Englishness' in design was identified too. A brochure produced to celebrate Waring and Gillow's contribution to the British Empire Exhibition at Wembley in 1924 drew attention to a number of specifically 'British' and 'English' traits, but significantly the terms were used interchangeably.[71] Titled *Past and Present*, this brochure underlined tradition and continuity, craftsmanship and artistic training, which resulted from 'this national passion for honest excellency, together with a certain richness of nature, and generosity of spirit'.[72] In the context of anxiety prompted by Britain's changing world role, the emphasis was on celebrating 'the greatest Empire the world has ever known', but alongside this was the familiar acknowledgement of the 'problems of a New Age'.[73] After describing the company's exhibit at Wembley, which included various displays based on Tudor, Jacobean and Georgian designs, it was suggested that 'The 1924 Bedroom' marked an attempt to

> express a modern tendency in decoration. It is expressive of life in 1924 and as charming and refined as the best of old . . . Its effect depends upon simplicity, straight lines, and simple masses, as well as beauty and richness of material plainly used.[74]

This design, indicative of Waring and Gillow's attempt to produce a modern style, not one based on historical precedents, showed an awareness and knowledge of French Art Deco. With its subtle decorative scheme of silver-grey walls, silver paper recesses, rich grey handmade curtains, grey sycamore furniture, deep mole-coloured carpet and pale yellow and orange lamps and lampshades, it introduced a very different European modernity to the British consumer. This European, and more specifically French style, was reinforced in 1928 by a company exhibition of *Modern Art in French and English Decoration and Furniture* in London. Writing in the foreword, Lord Waring reminded the visitor that the firm had played an important part in developing the character of English furniture from the late seventeenth century, but now recognized 'the modern movement in domestic art' and the role of the company in helping to shape this.[75] Criticizing the excessive individualism of what he termed 'the Continental school', he claimed the English 'are a people of precedent'.[76] The exhibition comprised 68 rooms – ten of which were French, designed by Paul Follot, director of the Modern Art department of the Waring and Gillow store on the Champs-Elysées in Paris, and ten English rooms, designed by Serge Chermayeff, director of the Modern Art department at the Oxford Street store in London. Both English and French displays consolidated the modern approach first seen at the British Empire Exhibition of 1924. The French dining room had a wrought-iron screen by the French designer Edgar Brandt; the walls were panelled with 'Duco', a new cellulose process, in shades of gold and brown to create an exotic scene; and the lighting was distinctly modern, being hidden in a central ceiling cornice. Equally the English Flat, which included a study, drawing room, double bedroom and young girl's boudoir, had a feeling of the 'Moderne style', but the materials were more traditional and restrained. The Moderne style was a synthesis of a number of styles and approaches, which taken together evoked modernity and engaged with aspects of modernist thinking and practice, even though individual elements might be drawn from the past and the present. Particularly 'modernist' was the interest in new materials and new technologies, the aim to develop a new visual language based on abstract, non-representational forms and simplified decoration, and the recognition that modern life required new modes of representation. Simple lines and striking colours including black and yellow were used in the drawing room with black wood furniture, whereas in the double bedroom 'Moderne' textiles and carpet by Alan Walton and Paul Follot combined with still-life paintings by Ben Nicholson and Christopher Wood to create a contemporary effect. Nevertheless, as Chermayeff astutely observed, the production of these designs was dependent on the highly skilled Waring & Gillow craftsmen, whose lineage went back to the eighteenth century, rather than modern industrial methods.[77]

The French Dining-room, Waring & Gillow exhibition in 1928, designed by Paul Follot.

The English Double-bedroom, Waring & Gillow exhibition in 1928, designed by Serge Chermayeff.

By the late 1920s, in addition to quality craftsmanship and materials, Waring & Gillow also catered for those with more modest incomes who nevertheless wanted well-made furniture – both period and modern. In a catalogue of 1927, *Suggestions for Spring Furnishings*, the consumer was offered 'the characteristic treatment of the famous period styles to the requirements of the present day conditions'.[78] As well as reproduction antique furniture made from old timbers, the company stocked a full range of household equipment, including china tableware such as Copeland Spode's 'Old Lowestoft', cretonnes and glazed chintzes, and an array of electrical appliances, including standard lamps, kettles, fires and irons. Provision of this type was usual, and a company such as William Whiteley Ltd of Queen's Road (now Queensway), London, aimed to offer a similar service to Waring and Gillow.

Whiteley's, offering lower prices combined with its 'system of easy payments', was advertised as 'the best in the world'.[79] Like Waring and Gillow, it offered complete household furnishings with estimates for a small two-bedroom house or flat for £175, and for a three-bedroom house (to include a maid's room) for £295. In the company's sales literature there was much to chose from, with hybrid ranges of inexpensive oak furniture based loosely on Tudor-Jacobean, Arts and Crafts, neo-Georgian and Moderne styles named after English towns and country houses, such as the 'Kenilworth' oak dining-room and the 'Wallingford' inlaid bedroom suite. Designs such as these were criticized by design reformers, who railed against what was considered to be the deception of reproduction. The DIA *Quarterly Journal*, for example, criticized designers who produced an old-world feeling in new houses, 'in order to catch the eye of the half-educated'.[80]

The DIA, which aimed to encourage 'a more intelligent demand amongst the public for what is best and soundest in design', grappled with the problem of how to improve public taste, encourage manufacturers to produce better design, and persuade the retailer to stock only good-quality products.[81] Writing at its foundation, the Scottish architect and designer Robert Lorimer, perhaps surprisingly, observed that 'the British Arts and Crafts Movement has had little effect on the general level of design of ordinary commercial goods'.[82] Like many, Lorimer admired the apparent efficiency of German design, attributing this to the close relationship between designer and industry, which he contrasted with the approach of the British Arts and Crafts Movement. He suggested that the latter had achieved little over the previous 30 years other than stimulating 'a comparatively small group of men and women to produce "applied art" . . . for garnishing the house beautiful'.[83] Craft ideas were, however, revived and reworked in the 1910s and '20s as designers headed back to the countryside to rediscover craft techniques, but significantly these were inspired by non-Western craft traditions as much if not more than the Arts and Crafts Movement. Most well known now was Bernard Leach, who established a pottery in St Ives in Cornwall in 1920 with the Japanese potter Shoji Hamada following his return from Japan and China.[84] But women were important contributors to the reinvigoration of the crafts and design as well. Empowered within the context of the suffrage movement, many women led independent lives, travelling, studying and ultimately establishing craft workshops in rural locations, gaining employment as architects and designers, or setting up independent design businesses. Carving out new roles for themselves, many were socially and economically privileged.[85] The hand-weaver Ethel Mairet drew on the colours and techniques that she had observed during her visits to Ceylon, Yugoslavia and Scandinavia in the 1900s for her textiles.[86] Her work marked a step change from the naturalistic forms of Arts and Crafts weaving in that it was abstract and experimental, and in her use of modern materials, Mairet showed a willingness to engage with the machine and modernity. The potters Katherine Pleydell-Bouverie and Norah Braden worked together at Coleshill in Wiltshire between 1928 and 1935, producing work inspired by Tang and Song wares, but also like Mairet, they were informed by modernist ideas. They, along with several others working in craft, believed that materials – whether old or new – had intrinsic qualities that demanded expression. They recognized that it was by means of a universal rather than a specific design language that contemporary life and experience were best expressed, and they searched for a new visual language that was abstract, non-representational and used decoration sparsely. Writing in 1930, Pleydell-Bouverie expressed it thus: 'I want my pots to make people think, not of the Chinese, but of things like pebbles and shells and birds' eggs and the stones over which moss grows.'[87] Women interior designers were important arbiters of taste in the upper echelons of society, and an understanding of pre-nineteenth-century styles was a prerequisite for those such as Sybil Colefax and Syrie Maugham, whose decorating schemes were reported in *Vogue* as well as *The Studio*. The taste for neo-Georgian and Regency provided a form of aesthetic purification, perhaps helping to create the preconditions for the more abstract modernism of the mid-1930s; as Osbert Lancaster put it at the end of the 1930s: 'Today, the more sensible of modern architects realize that the desperate attempt to find a contemporary style can only succeed if the search starts off at the point at which Soane left off.'[88]

Throughout the 1920s the DIA took a strong line against historical revivalism, the anti-machine rhetoric of the Arts and Crafts Movement, and elitist design practices, which focused on the privileged and ignored the everyday. But, significantly, the ideas of group members had been shaped within these very contexts. With groups in Northampton, Manchester, Bristol and Bradford as well as London, the DIA tried to invoke a campaigning approach to persuade and cajole all those involved in design to improve its quality at an accessible price. Its chairman was the writer John Gloag, and members included the furniture retailer Ambrose Heal, the designer

Stoneware bowl with fluted sides by Katherine Pleydell-Bouverie, 1930s. Hawthorn ash glaze in grey-green.

Harold Stabler, the architect Charles Holden and patrons such as the manufacturer W. J. Bassett-Lowke, who were all in positions of influence. In 1918 the DIA member and designer Hamilton Smith, addressing Stoke-on-Trent potters on 'The Ideals of the Design and Industries Association', tried to encourage the pottery manufacturers to turn away from historical revivals: 'The pottery of our time seldom expresses anything in particular, and is to a great extent copied from things produced in past ages', he said. Arguing that the DIA had no essential quarrel with reproductions, he nevertheless suggested: 'The present-day designer is being stultified . . . denied his legitimate right of expression and becomes a mere mechanical drudge.'[89] With a broadly educative slant, the DIA organized visits at home and abroad; it reviewed exhibitions such as one promoting modern furniture at Shoolbred's store in London in 1928 and the *North East Coast* exhibition in Newcastle in 1929, and via the DIA *Quarterly Journal* it promoted discussion on many aspects of design. It attacked retailers such as Harrod's and Waring & Gillow, arguing that

> the world in general is growing very, very bored with the antique (which is quite out of keeping with materials and conditions of to-day) and that firms and nations who do not give their artists a chance will wake up to find them-selves out of the running.[90]

Drawing an analogy with fashion in a bid to encourage contemporary design rather than reproductions, the writer argued that women's fashions in the 1920s were simple and practical in response to the conditions of modern life, since 'she now has to stride out in step with men'.[91] A keen observer at the time might have recognized that women's fashion had had its own 'neo-Georgian' phase in the first decades of the twentieth century. Then Paul Poiret and his followers had encouraged the stripping away of Victorian and Edwardian excesses by going back to the late eighteenth-century styles and by rediscovering dress from non-Western cultures in a manner not dissimilar to that of the disparaged furniture retailers, as well as earlier modernists.

Towards the end of the 1920s DIA thinking about modernity began to be couched in the language of Continental modernism as the organization focused on the need to design well for economic growth and as new members such as Jack Pritchard and Frank Pick, who became chairman in 1932, were ascendant. This point had particular currency towards the end of 1929 as recession turned to depression. In an article entitled 'Advertising the Nation's Workshop' in the *DIA Quarterly Journal* of July 1929 (just prior to the Wall Street Crash in October 1929), the *North East Coast* exhibition in Newcastle upon Tyne was discussed in economic terms:

> The exhibition is the outcome of a co-operative effort by the Tyneside industries to conquer the trade depression which has fallen so heavily on this district which used to be known as the 'nation's workshop'.[92]

Rationalism and functionalism harnessed to address Britain's economic problems were elements of this revised DIA thinking, which, emerging towards the end of the 1920s, coalesced with the agenda of those influenced

Empire Marketing Board stand at *North East Coast* exhibition, Newcastle upon Tyne, 1929.

Designing Modern Britain

Aitken & Rush, radio retailers' stand at the North East Coast exhibition, Newcastle upon Tyne, 1929.

by Continental modernists such as Pritchard and Pick, but also Bauhaus émigrés who arrived in the mid-1930s.

But as this chapter has shown, although increasingly influential, this new emphasis on rationalism and functionalism represented only one strand of Britain's engagement with modernity and modernism in the 1910s and '20s. Discussing the forcible exile of Omega and Bloomsbury from definitions of modernism, Christopher Reed argued that it is in these that the complexities of British responses to modernism were evident.[93] Equally in critical writing (in the 1930s and after) on various subjects: Omega, the 'Moderne' style, the use of craft and attitudes towards decoration and tradition in design, the preoccupations, priorities and prejudices of an increasingly prescriptive and hegemonic Continental modernism can be observed. Fry was a hugely influential writer and thinker. His implicit questioning of artistic hierarchies and enthusiasm for the visual culture of non-Western countries linked back to the Arts and Crafts Movement, but it also resonated with the ideas of Continental modernist theorists and practitioners, including the Swiss architect Le Corbusier and the former director of the German Bauhaus, Walter Gropius (one of the émigrés arriving in Britain in the mid-1930s). By the late 1920s the ideas of the latter had already begun to gain a foothold in Britain via periodicals such as *The Studio* and the *Architectural Review*, and significantly, as these developed between 1930 and 1950, they took on a distinctive shape. In the rhetoric of the time, there was a strong emphasis on utility, fitness for purpose and 'form follows function'. Decoration and historical styles became anathema, along with fashion and the transient, which were rejected in favour of the universal and the timeless. As a consequence, at the very moment that modernist ideas from the Continent were gaining ground, alternative modernisms already evident in British design began to be attacked or effaced from critical discourses. Thus craft, decoration and eclecticism, integral to modernist practices in Britain before 1930, were estranged after it.

3

'Going Modern, but Staying British': Design and Modernisms, 1930 to 1950[1]

Asking whether it was possible to go modern and remain British, the artist and designer Paul Nash pinpointed the growing dilemma for those who attempted to reconcile the internationalist tendencies of modernism with a respect for national qualities. A growing commitment to European modernist architecture and design coexisted and overlapped with a number of different approaches; thus modern meant not just modernist, Art Deco and Moderne, but also, perhaps surprisingly, some traditional styles. Increasingly articulated in modernist theory, order, structure and planning were also integral to past styles – in particular Georgian architecture and design – thus providing a sense of continuity between past and present.[2] To design reformers in Britain, being modern and being traditional were not necessarily opposites, but part of a continuum, and for retailers, builders, designers and consumers a synthesis of traditional and contemporary themes and styles became essential. It has been argued that a concern for 'Englishness' and tradition represented 'a deferral of modernity'; but this chapter proposes that modernity was not so much deferred as renegotiated in a number of ways.[3] Discussing British art of the 1920s, Charles Harrison suggested in 1978 that there was nothing innovative and progressive in Britain after the First World War: 'the twenties had been quiet years. There were no very challenging exhibitions, no invasions by outlandish foreigners. No significant groups were formed, no radical theories expounded.'[4] A similar line was taken in the catalogue that accompanied the exhibition *Modern Britain, 1929–1939*, held at the Design Museum in 1999; in this, the architect Norman Foster, after attributing almost all crucial modernist buildings of the 1930s to émigré architects (ex-Bauhaus staff Walter Gropius, Marcel Breuer and László Moholy-Nagy arrived in Britain in the mid-1930s), went so far as to argue that modernism 'only arrived in Britain

Wireless in Bakelite designed by Wells Coates for EKCO (E. K. Cole Ltd) in 1932, manufactured in 1934.

with these émigrés'.[5] As we saw in chapter Two, this was not the case. A variety of responses to modernity emerged in the 1910s and '20s; and although these were later reconfigured, most included elements of what would be described as modernist practice – a concern with new materials and innovative technologies; a desire for a radical visual language based on abstract, non-representational forms and simplified decoration; an engagement with universal design qualities rather than specific ones; and an awareness that modern life required new modes of representation. Designers, manufacturers, critics and consumers were therefore already 'going modern and being British' before the arrival of the European émigrés. For Norman Foster, modernism in Britain in the 1930s amounted to a handful of buildings conceived and erected against the odds by 'pioneers'. These were without exception based in the south-east of England, if not London itself. Equally, in this narrative, modernism somehow bypassed Scotland, Ireland, Wales and the 'English' regions.

No doubt persuasive, since it locates design in the hands of those few who produced one-off avant-garde objects, the notion of design as the province of 'pioneers' has limitations as a way of thinking about everyday design. Pioneers have had a special place in modernist histories of design, as curators, critics and historians have been swayed by the notion of outstanding individuals battling against the odds.[6] At first glance, the designer and architect Wells Coates was one of these. He designed standardized units using modern materials in 1929, and although he was not termed a modernist at the time, his designs were unquestionably influenced by the complexity of modernist practices and theories. A member of Britain's fledgling avant-garde, Coates was informed by modernist ideas to a greater or lesser degree, as were several others – the artist/designer Paul Nash and the ceramic designer Susie Cooper, for example – who were part of a matrix linking social, artistic and educational networks. Indeed, they were embedded and located within very specific social and cultural milieux, rather than being outsiders. This was evident at the time: in the commissions they gained (Coates's patrons included Tom Heron, father of the artist Patrick Heron; George Strauss, Labour MP for Lambeth North; and Jack and Molly Pritchard, Cambridge graduates, respectively an engineer and a bacteriologist); the articles they wrote (Coates, for example, wrote in *The Listener*, the *Architectural Review* and the *Architects' Journal*); the exhibitions to which they contributed (*British Industrial Art in Relation to the Home* in 1933, *British Art in Industry* in 1935); the groups they formed (Unit One, MARS); and in the organizations they joined (the Society of Industrial Artists). Coates and Nash were part of a middle-class cultural intelligentsia centred on London, whereas Cooper was based in Stoke-on-Trent. University- or art-school-educated, they were not necessarily wealthy, but they had contacts and patrons who were; thus their first designs were often either privately commissioned or the product of personal contacts (Susie Cooper's first job, for example, came through the efforts of the educator and designer Gordon Forsyth).

To some of the critics and writers who were Continental modernism's early supporters, 'British' reconfigurations seemed timid and lacking in conviction. But if we accept that modernism was not singular, but plural – that it developed through a myriad of interlinked factors, rather than individual inspiration – it is apparent that modernist practices in Britain were subtle and complex, and in design terms its origins lay in the 1920s, if not the 1910s. Certainly, there were art schools in Britain in the 1920s – driven in part by economic considerations, but also by aesthetic ones – that were 'modernist' in approach if not in name, and they were not all based in London. The Potteries Art Schools in Stoke-on-Trent were an example of the ways in which manufacturers, trade unions and educators came together following recognition that design in manufacturing industry needed to be tailored to the needs of the modern world.[7] The head from 1920 was the potter Gordon Forsyth, who set about 'reorganising the existing art schools on purely industrial lines. Up to that time people had been rather inclined to regard the art schools of the Potteries as being much more concerned with the fine arts than industrial requirements'.[8] By 1925 he had established a Junior Art Department, which aimed to raise the standards of apprenticeships by producing artistically educated men and women suitable for employment in the pottery industry, and by 1930 there were some 1,100–1,200 students attending the Potteries Art Schools.[9] Forsyth's own aesthetic lay in the Arts and Crafts Movement, but that did not inhibit him from recognizing the necessity for a new type of designer geared to the needs of industry. Such initiatives in Stoke-on-Trent were contemporary with those in Europe that led, for example, to the foundation of archetypal modernist institutions, such as the Bauhaus in Germany and the Vkhutemas in the Soviet Union, established in 1919 and 1922 respectively, but they were more pragmatic, less driven by experimentation and utopianism. Nevertheless, the curriculum at the Burslem School of Art was based on the idea that students had to have an understanding of industry and art, and art meant design, form and decoration rather than merely knowledge of styles – either past or present. Forsyth's Arts and Crafts background predisposed him to awareness that form and decoration should be integrated. But also Forsyth understood the practicalities of design for industry and the necessity of producing designers capable of industrial design. Importantly, these ideas were part of a wider concern in Britain to design for the needs of the modern world, and such views – with nineteenth-century origins –

formed a basis for new ideas in design education at Britain's foremost art school, the Royal College of Art, in 1946–7. With close connections to pottery manufacturing, Forsyth established links with progressive manufacturers such as Josiah Wedgwood and Sons Ltd, A. J. Wilkinson and Foley China through networks focused on organizations such as the DIA and the Society of Industrial Artists (SIA). He was also acutely aware of the importance of the consumer in the good design equation. Consumption of domestic pottery in this period was governed by a host of factors: smaller modern houses required new sets of tableware; a larger middle-class market had more choice of where to buy, what to buy and how to pay; and, of course, the Second World War. Companies such as Josiah Wedgwood had begun to change before the First World War, but it was in the 1920s that this gathered momentum. As we have seen in the 1920s, 'the modern' was represented by hand-painted mass-produced pottery, such as that designed by Louise and Alfred Powell and Millie Taplin. Based on shapes that had their origins in craft forms and eighteenth-century prototypes, these were nonetheless made by industrial methods and decorated with simplified abstracted patterns and colours that revealed the impact of modernist art, which reached the pages of the trade as well as art journals. In addition, sophisticated advertising campaigns coordinated by companies such as Shelley Potteries, Josiah Wedgwood, A. J. Wilkinson and Susie Cooper were highly effective in persuading the consumer that modernity was desirable and compatible with tradition and continuity. Consumers were keen to be modern, but they were also interested in traditional design values, hence the success of a company such as Wedgwood in combining modern decoration and form with eighteenth-century prototypes. Wedgwood, for example, advertised its 'Living Tradition', whereas Shelley used the youthful 'flapper' Elsie Harding to promote the modernity of its wares.

Synthesizing and mediating both tradition and modernity, the home became a particular focus for design reformers between 1930 and 1950 as they attempted to educate the public in good design and taste. Significantly, the home could be modern, modernist and 'English' at the same time. It was constructed using standardized parts, often incorporating new materials and technology – metal-framed windows, wired for electricity and with a garage – but it might also look countrified and 'English' with tile-hanging, half-timbering, over-hanging eaves and bay or oriel windows. Located in the suburbs or the post-war New Towns, and connected to the towns and cities with arterial roads, trams and tubes, this was a far cry from an idealized rural idyll, but stylistically it looked back to vernacular styles as well as forward to modern ones.

An engagement in the modern was articulated in design throughout the middle decades of the century, but it did not result in one coherent set of theories and practices. Instead, artists, designers and architects grappled with 'English' and 'British' crafts and traditions alongside learning about new technologies and their myriad applications. But at the same time they were interested in crafts and traditions from other countries – from the 'Empire', but also from less well-documented countries nearer home. Thus the textile designer Ethel Mairet, based in her workshop Gospels in Ditchling, East Sussex, visited Ceylon, Scandinavia and Yugoslavia researching indigenous dyeing and weaving techniques. She attempted to reconcile hand-weaving to the needs of modern life by designing for the machine, using new materials alongside traditional ones.[10] Mairet was one of a number of designers/craft makers (others included Phyllis Barron and

Chestnut and Vernede, two hand-block printed textiles designed by Phyllis Barron and Dorothy Larcher, 1920s–'40s.

Dorothy Larcher) whose work in the 1930s and '40s had a distinctive modern feel that exemplified important aspects of modernism.[11]

Barron and Larcher were designers and makers of hand-block printed textiles. They trained at the Slade School of Art and Hornsey School of Art respectively before entering into partnership in 1923. Barron was already established as a textile designer by this time; she had had a huge commission for the Duke of Westminster's yacht, *The Flying Cloud*, bringing other potential clients. Larcher, in contrast, had recently returned from India, where she had researched indigenous methods of dyeing and printing textiles. By 1930 the two women had established their collaborative practice. They used cotton, silk, linen, velvet and organdie printing with hand-cut wood blocks in one or two colours. Dyes were thickened with gum to produce a mottled, uneven surface at odds with standards of finish in commercially printed textiles. Exhibiting consistently from the end of the 1920s through the 1930s, particularly at Muriel Rose's Little Gallery in Ellis Street, London, the patterns they designed such as Chestnut and Vernede were remarkably abstract. Based on craft techniques, the visual style of their work connected with the formal simplicity of modernism, yet the hand-made surface of the prints – most evident in the mottling – was a far cry from the technological rationalism associated with Continental modernists such as Marcel Breuer and Le Corbusier. It was, however, modernist in orientation, challenging preconceived ideas about the nature of 'design'. Something akin to this concern for surface – texture, colour, technique – could be found in the work of studio potters of the time, such as Katharine Pleydell-Bouverie and Norah Braden. Undermining accepted ideas about 'design' in textiles and ceramics, this concern for 'surface' and 'form' did not originate in 'English' craft traditions – it owed more to the East, both in spirit and detail – but traditional 'English' culture was, nevertheless, 'indexed in these uneven surfaces'.[12] Thus painterly modernism, craft and non-European cultures intersected, adding another layer to what might constitute modern design in 1930s Britain.

The Condition of Britain

Writing on 'the condition of Britain in the '30s', Charles Loch Mowat discerned an increased introspection, as 'the country turned inward, and concerned itself more with its own ills than with the cares of the world'.[13] This prompted more reflection and analysis about the state of the nation (perhaps the most novel and extensive of these was the Mass-Observation survey at the end of the decade). But Mowat also noted an increased social consciousness that led to a growing political awareness evident, for example, in attitudes to the Spanish Civil War and the rise of fascism. For all its intro-spection in some areas of British cultural life, there was a measured interest in the 'new', necessitating an outward-looking stance. In parallel was the idea – most powerfully represented in J. B. Priestley's *English Journey* of 1934 – that 1930s Britain was at least two 'Britains'.

At the end of this famous journey across England, Priestley concluded that there were many Englands, but perhaps three stood out. There was 'Old England', 'the country of the cathedrals and minsters and manor houses and inns, of Parson and Squire; guidebook and quaint highways and byways'.[14] There was also nineteenth-century England, 'the industrial England of coal, iron, steel, cotton, wool, railways; of thousands of rows of little houses all alike'.[15] Finally, there was the third England, 'the new post-war England, belonging far more to the age itself than to this particular island. America, I suppose, was its real birthplace.'[16] Of the first 'England', Priestley wrote:

> we all know this England, which at its best cannot be improved upon in this world . . . It has long ceased to make its own living. I am for scrupu-lously preserving the most enchanting bits of it, such as the cathedrals and the Cotswolds, and for letting the rest take its chance.[17]

Of the second 'England', he pointed out: 'this England makes up the larger part of the Midlands and the North and exists everywhere; but it is not being added to and has no new life poured into it'.[18] After a depressing account of its shortcomings, Priestley speculated as to whether the inhabi-tants of this England were any better off than those in the pre-industrial one, before concluding: 'they all rushed into the towns and the mills as soon as they could, as we know, which suggests that the dear old quaint England they were escaping from could not have been very satisfying'.[19] He described his third 'England' derisorily as 'a large-scale, mass-production job, with cut prices'.[20] It was as 'near to a classless society as we have got yet. Unfortunately, it is too cheap.'[21] It's being too cheap – implying fake – he attributed to the influence of America, which among other things had brought:

> arterial and by-pass roads . . . filling stations and factories that look like exhibition buildings . . . giant cinemas and dance-halls and cafes, bungalows with tiny garages, cocktail bars, Woolworths, motor-coaches, wireless, hiking, factory girls looking like actresses, greyhound racing and dirt tracks, swimming pools, and everything given away for cigarette coupons.[22]

This new post-war England was standardized and regimented, even though it might be a cleaner and healthier place than that of nineteenth-century industrialism. Priestley argued that these three 'Englands' were mingled together in every part of the country, but some areas fared better than others, particularly the south. As he put it,

> was Jarrow still in England or not? Had we exiled Lancashire and the North-east coast? Were we no longer on speaking terms with cotton weavers and miners and platers and riveters? If Germans had been threatening these towns instead of Want, Disease, Hopelessness, Misery, something would have been done quickly enough.[23]

Writing some 15 years later in the introduction to a new edition of *An English Journey*, Priestley observed that not only had perceptions of the book changed (it had been received initially as social commentary), but England itself had changed, largely because of the Second World War. The social injustices highlighted in his book were now the target of the Labour Government's post-war social reforms.

The British economy, however, was in disarray at the end of the Second World War; although the 1930s had witnessed growth in domestic consumption, this 'was largely a middle-class phenomenon. The exceptions to this general rule were the radio, the vacuum cleaner and the iron.'[24] A result of regional inequalities, ownership of these consumer durables in the 1930s was concentrated in the wealthier parts of Britain; for example, consumption of electricity by domestic consumers in the south-east was more than twice that in the north-east of England.[25] Middle-class women consumers in particular were addressed by magazine and newspaper advertising, but surprisingly these new goods proved to extend their time spent on housework: 'domestic technology eased the reallocation of housework away from the domestic servant to the middle-class housewife and the occasional help'.[26] In addition, standards of domestic hygiene were raised as women were persuaded that housework was a 'profession' that demanded specific new skills; organizational, technical and managerial. Writing in *The Electrical Handbook for Women* in 1936, its president, Margaret Moir, and director, Caroline Haslett, proposed 'a new technique of Home Management, an alliance of Domestic craft with Engineering'.[27] Describing the cook who used an electric cooker as a technician, she observed: 'statistics show that there are well over a quarter of a million cookers on hire in this country, whilst others, unrecorded, are owned by consumers'.[28]

Consumption of domestic goods in the 1930s was closely linked to the provision of new homes, particularly owner-occupier housing, but because

Designing Modern Britain

of central and local government subsidies, this also included council-house building. Four million dwellings (both local authority and private) were built in the interwar years, and a quarter of them replaced slums. The shift to owner occupation continued, with 20 per cent owning their homes in 1939 as opposed to 5 per cent in the mid-nineteenth century.[29] Elizabeth Roberts interviewed one such family who moved from a two-up, two-down rented terrace to a newly-built house nearby in 1937. With a bathroom, electricity, a gas cooker and hot-water supply, this cost £295.[30] Although prosperity increased for some members of the working class, there was real hardship for others during the first half of the 1930s because of widespread unemployment in those parts of the country dependent on heavy industries: coal, shipbuilding, engineering, and iron and steel. In towns such as Jarrow and Hebburn on Tyneside

> there was nothing in the whole place worth a five-pound note. It looked as much like an ordinary town of that size as a dustbin looks like a drawing room. Here again, idle men – and not unemployable casual labourers but skilled men – hung about the streets, waiting for Doomsday.[31]

Jarrow's unemployed famously marched to London in 1936 to protest that 80 per cent of its workforce was unemployed, but by this date the threat of war had in part stimulated the economy and recovery was on the way. The underlying structural problems in the economies of regional heavy industries remained, however, to resurface after 1945.

By 1939 a war economy was created as government powers increased. The main outcome of this was

> that the market oriented economy of the interwar years was replaced by a centrally managed economy in which the state allocated the most important resources, decided what should be produced, and determined how much should be paid for it.[32]

This, as we shall see, had enormous repercussions for design, because central government took control of materials, factories and labour. From 1941 all types of goods – clothing, furniture, food and consumer goods – began to be rationed using a points system. To a large extent, the government managed consumer demand on the basis of need rather than desire; this was an essential economic strategy in order to show fairness in the distribution of goods.[33] Surprisingly, food rationing lasted into the mid-1950s, but clothing ceased to be controlled in 1949.

Although the British economy was in very poor shape in 1945, fairly rapid recovery was achieved in the decade after the war, partly by the maintenance of a managed economy by the new Labour government. Wartime restrictions had created an insatiable demand for consumer goods, but the continuation of rationing caused the period of austerity to last until the late 1940s, because the emphasis lay on getting industry into full production to produce exports. Domestic consumers, it seemed, were the last in line to buy new goods; instead, these were targeted for export, as was apparent at the *Britain Can Make It* exhibition of 1946 at the Victoria and Albert Museum. State management of the economy was paralleled by increasing state intervention in design; significantly, modernist theories and practices, perceived as foreign and radical in the 1930s, were deployed to represent the 'brave new world' of post-war Britain. Inevitably, this too was 'managed' by the perpetuation of state involvement in design policy and education, exemplified by the formation of the Council of Industrial Design in 1944. The conjunction of modernist aesthetics with centralized government planning contributed to the reinforcement and consolidation of modernist ideals at a crucial historical moment, from the mid-1940s to the early 1950s. These essentially elitist design practices were undermined from within and without, however, as questions of 'national' identities preoccupied many of those engaged in building a better Britain. Design was increasingly international and in Britain it was influenced by European ideas – from Scandinavia, Italy and Germany – as well as those from the economically strong USA from 1930 to the 1950s.

Modernist Designers: Paul Nash, Wells Coates and Susie Cooper

A sense of urgency informed debates from the mid- to late 1920s regarding the designer's role in developing new products for modern life. Increasingly, these took place within a framework of modernist ideas and practices; some were already evident in Britain as we saw in chapter Two, but others originated in Europe and the USA. Working in Britain during this period for Waring and Gillow, the modernist designer and architect Serge Chermayeff disclaimed all knowledge of the USA, though others were well aware of American developments, particularly the application of the latest technologies and the deployment of innovative commercial strategies.[34] The exchange of ideas went both ways because the USA remained a crucial export market for British goods, but French Art Deco proved particularly influential from the mid-1920s and European modernist writings (published in *The Studio* and *Architectural Review* from the late 1920s) proposed

Designing Modern Britain

radical solutions to the problems of designing for contemporary life. In different ways, the designs of Paul Nash (1889–1946), Wells Coates (1895–1958) and Susie Cooper (1902–1995) exemplified the variety of modernist ideas and practices in Britain at this time.

Paul Nash's design practice is particularly revealing of the ways in which an artist worked in design between 1910 and *c.* 1935. Nash, who initially trained in illustration, switched to art, enrolling at the Slade School of

Poster advertising an exhibition of war paintings and drawings at the Leicester Gallery, London, 1918, designed by Paul Nash (lithograph).

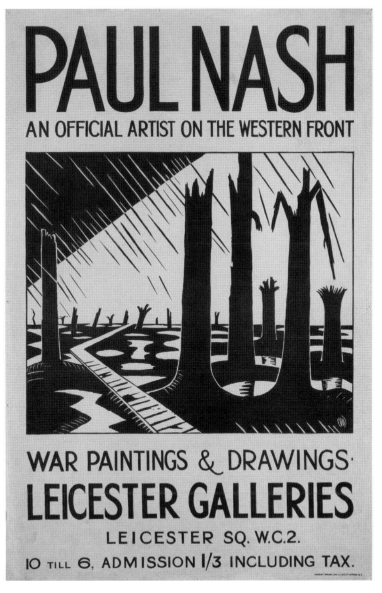

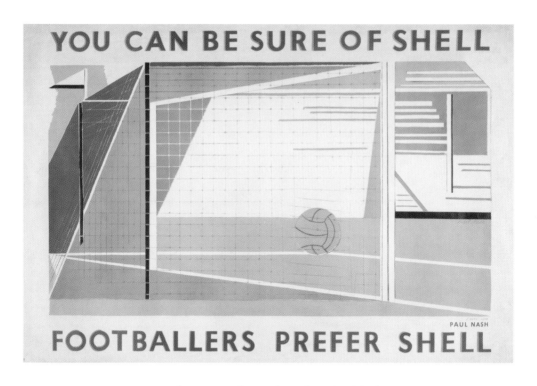

YOU CAN BE SURE OF SHELL

PAUL NASH

FOOTBALLERS PREFER SHELL

Art in 1910. Retaining an involvement in design throughout his life, he exhibited paintings at the New England Art Club exhibition in 1913, where he attracted the attention of Roger Fry. After gaining further critical notice as a war artist, he designed book illustration and textiles from 1918. Both owed a good deal to the Arts and Crafts Movement, particularly in terms of technique (the use of wood engraving and block printing), but also in terms of style. In fact, his earliest bookplates from 1910 referenced medievalism via William Morris's Kelmscott Press, although by the mid-1920s his designs for textiles and book jackets showed an awareness of French Art Deco and the Moderne styles that began to influence design in Britain after 1925. To some extent Nash was a transitional figure working at the end of the Arts and Crafts Movement, but responsive to the ideas of Fry and associated with Omega Workshops in the 1910s. He remained in touch with various craft networks throughout the 1920s as a result of his textile designs – he designed for Celladine Kennington's Footprints company (founded in 1925) and for Elspeth Little's Modern Textiles in Beauchamp Place (set up in 1926). Although he taught design at the Royal College of Art in the mid-1920s and again at the end of the 1930s, he saw himself primarily as a painter, but, typical of the period, he moved from one to the other with relative ease. Nash became particularly focused on design in the late 1920s and

Poster designed by Paul Nash for Shell-Mex, 1935 (colour lithograph).

Designing Modern Britain

early '30s, when he produced designs for a number of manufacturers and clients: the publishers Curwen Press, Chatto & Windus and Faber and Faber; the textile manufacturers Cresta Silks Ltd, Old Bleach Linen Co. and Footprints; posters for Imperial Airways, Shell-Mex Ltd, BP Ltd and London Passenger Transport Board (as well as moquette fabrics for Underground and bus seat covers); rugs for the Edinburgh Weavers; glass for Stuart and Sons; and ceramic tablewares for A. J. Wilkinson and E. Brain & Co. Notwithstanding his reputation as a painter, Nash's practice exemplified that of a freelance industrial designer.[35] To a certain extent this involvement in design was in response to the economic conditions of the early 1930s, since work was hard to find (the architect Keith Murray also worked as a designer for similar reasons). But artists and architects, who had, after all, considerable skills in design as a result of their training, also responded to the call to improve the quality of design in industrial products. In part, this was caused by Britain's precarious export position, but also by modernist ideas about the importance of designing for industry and the important role of the abstract artist in this. Nash was a member, later becoming President of the Society of Industrial Artists (founded in 1930, this was concerned with the professionalization of design), the Council for Art and Industry, and he exhibited at the important *British Art in Industry* exhibition at the Royal Academy in 1935. In 1933 he founded Unit One, a group of architects, designers and artists (of which Coates was another member), aiming to express 'a truly contemporary spirit, for that thing which is recognised as peculiarly of to-day'.[36] Unit One was an example of the avant-garde collaborations between artists, architects and designers typical of modernism, but Nash retained a connection with Arts and Crafts principles; as Lambert argued, 'his feeling for the essence of nature is as clear in his decorations for breakfast sets as in his canvases'.[37] Nash believed that the 'English' tradition of design was located firmly in the eighteenth century, recognizing, along with many of his contemporaries, 'modernity' in its simplicity. Nash's design work tailed off towards the mid-1930s, just as émigré modernist architects and designers arrived in Britain from Europe. He was not enraptured by the abstract, rectilinear aesthetic proposed by such men as Herbert Read and Walter Gropius, and for all his promotion of good design in industry he remained an 'artist and an individualist'.[38] Nevertheless, Nash was part of the matrix of modernist design practices: organizing, campaigning, publishing and designing. His path, which included both art and design from the early 1910s to the mid-1930s, represented a negotiation of various forms of modernist practice; Fry and Omega, craft, Art Deco and the Moderne, and European modernism. While clearly modern, his work retained an interest in tradition and decoration.

Wells Coates was a very different designer to Nash, although there were some shared interests. His article 'Response to Tradition', published in *The Architectural Review* in November 1932, contributed to the increasingly contested field of what constituted modernism. His proposals refused the compromise that many believed to be characteristic of earlier interpretations of modernism in Britain. Coates's didactic views were an affront to those such as the DIA chairman John Gloag, who promoted a synthesis of 'Englishness' and modernism. Aiming directly at the DIA and kindred reformers, Coates set about demolishing their belief that it was possible to use design elements from the past in the present:

> These societies for the preservation of this, the conservation of that, who say to the commoners: 'You must not erect your sham Tudor tea-shop, your sham Greek details all over your petrol station . . . ' all this is based on a completely wrong psychology. For *you* have debased the great traditions. *You* have converted a Greek temple into a banking-house; *you* have plastered the second-hand columns of the ancients on to the grocers' shops of Oxford Street. The ugly petrol station is the logical conclusion of your efforts.[39]

To jolt these reformers from their complacency, he proposed taking as a guide, 'a stranger to the West, one born and brought up according to the inflexible customs of an ancient civilisation of the East'.[40] Coates, who was born in Japan and lived there until he was 18 years old, tells us that his imaginary guide has travelled to Europe, but has been told that 'a man whose eyes have been trained in the East will only rarely want to open them in the West'.[41] To overcome this, in order to see beneath the 'confusion of appearances and re-appearances, the accretion of layer upon convoluted layer of architectural growth', he provided himself with a kind of aesthetic x-ray, 'to track down its underlying shape, the sources of its traditions'.[42] In this essay, Coates's assimilation of the ideas of Walter Gropius and Le Corbusier was apparent: 'it is for architects to invent, and to exhibit, a new architecture which will quite naturally be accepted and demanded by the people'.[43] He had a rather different take on modernism than some of its other followers, however. He was a vehement critic of those who believed that modernism was merely about functionalism, believing that 'every change in conditions brings with it new possibilities of systems of impulses, needs, expectations, attitudes'.[44] Coates was a perceptive thinker and a talented industrial designer; trained in mechanical and structural engineering at McGill and British Columbia universities in Canada, followed by a PhD in engineering at London University, he worked as a designer and architect

Minimum Flat for Lawn Road Flats, Hampstead, London. Exhibited at the *British Industrial Art in Relation to the Home* exhibition at Dorland Hall, 1933, designed by Wells Coates.

until his death in 1958.[45] He began designing shop fronts, interiors and fittings in 1928, but rapidly expanded his activities to include furniture, architecture, exhibition stands, recording studios, radios, aircraft, sailing craft and exhibition design. Underpinning most of his designs was a commitment to using new materials – concrete, steel, plastics and plywood – an enthusiasm for innovatory design solutions and a concern for abstract forms. In his designs for Cresta Silks shop fronts, the brief was to design adaptable, inexpensive units. This theme of standardized units or modules was a feature of European modernism and it re-emerged in his work for Isokon, the company set up by Jack Pritchard in 1930 to produce furniture and housing. For Isokon, Coates designed furniture, interior fittings and housing, including Lawn Road Flats in Hampstead, London, in 1933–4. There were 22 'minimum' flats, which marked a clear response to modernist practice in Europe, where 'minimum space' was an integral element of the social housing schemes developed by city authorities to house those in need. But once inside the Lawn Road minimum flat, it was clear that this was not mass housing for the working class. Exhibited initially as a prototype at the *British Industrial Art in Relation to the Home* exhibition of 1933, the minimum flat had a plethora of modern conveniences: electric cooker, refrigerator, radio and central heating designed to suit the young middle-class professional who required services, not things, and freedom 'from enslaving and toilsome encumbrances in the equipment of the modern dwelling scene'.[46] As Coates wrote, 'the home is no longer a permanent place from one generation to another', and it was obvious that the Lawn

Road flats were to be equipped for a new type of person.[47] With a maid service, central kitchen and, by 1937, a restaurant, its first inhabitants included notable figures such as the crime novelist Agatha Christie, the architect Arthur Korn, the writer and journalist Lance Sieveking and the émigré architects Walter Gropius and Marcel Breuer.

Like Nash, Coates's practice as a designer was extensive. In 1932 he worked with Raymond McGrath and Serge Chermayeff, designing interior and technical equipment at the new BBC broadcasting studios in London before designing studios in Newcastle upon Tyne and Manchester (since demolished). In these he did detailed technical equipment and fittings, including dramatic effects studios, control rooms, gramophone studios and equipment such as microphones.[48] The interiors in Newcastle were typical of his designs: simple, geometric and abstract with subtle, restful colour – green, grey, black and beige – and using new technologies and materials, including plywood, Bakelite, tubular steel, though combined with ebonized hardwoods for detailing. The overall look was modernist, but at the same time subtly decorative, referencing the Moderne style. In the same year, 1932, Wells Coates won a competition organized by EKCO, the manufacturer of Bakelite products for the design of a radio set. Astutely, he identified the

Ceramic tablewares designed by Susie Cooper in 1934. Kestrel shapes with Crayon Lines pattern.

Designing Modern Britain

Susie Cooper Pottery showroom in Woburn Place, London, mid-1930s. Note the presentation and display of the ceramics.

nub of the problem inherent in radio design at the time, 'a radio should never be distinguished as something else. It has its own important function in the home and is in many cases a possession regarded more as the indoor equivalent of a car than a piece of furniture.'[49] For this piece of portable equipment, Coates designed a relatively compact circular object that exploited the unique moulding qualities of Bakelite and required few moulding tools. This extremely modern design encapsulated the complexities of modernism in 1930s Britain – abstract, mechanistic and dependent on technological innovation – but nevertheless produced in a fake walnut burr Bakelite as well as a sleek black version in response to consumer demand. Coates, while unquestionably knowledgeable and committed to the tenets of European modernism, was vehemently anti-functionalist, as his designs for the BBC studios and EKCO amply demonstrate. His awareness of Japanese culture infused his work with a concern for the spiritual dimensions of architecture and design that transcended functionalism.

The ceramic designer Susie Cooper represented a somewhat different engagement with modernist practices in Britain in the early 1930s. She trained at Burslem School of Art, in the north-west of England, under the tutelage of Gordon Forsyth in the mid-1920s. Because of the nature of ceramic manufacture in Stoke-on-Trent and the demands of the market, her designs represented an ongoing engagement with decoration at a time when modernist critics were promoting minimal or no decoration.[50] Although constantly castigated by modernist critics, the pottery manufacturers in Stoke-on-Trent remained committed to decorated pottery. For those designers like Cooper who were interested in modernist design, the challenge was to develop a response to modernism that recognized the significance of decoration to both manufacturer and consumer. Like Coates, she understood that modern life required new designs, 'the drastic changes that have come over the domestic life of many people warrant the provision of smaller and better balanced services'.[51] To this end she developed new ranges of wares that were attuned to the changing function of tableware within the middle-class home. Described as 'a lady who designs from the standpoint of the lady', she implicitly recognized the importance of the female consumer.[52] European theorists articulated modernism as masculine, the result of science and technology, rationalism and standardization, and collapsed its negatives – decoration and fashion – into the realm of the feminine.[53] Decoration, however, was integral to earlier forms of modernism in Britain, and it had not been entirely abandoned by European exponents. Several designers (both male and female) employed a decorative language of subtle colours or neutral tones of cream, brown and black; they adopted a light, loose graphic touch and developed patterns that, although abstracted, were still recognizably drawn from nature. Pottery decoration tended to be small scale, often based on flowers, although most designers also produced patterns that were abstract and/or geometric, for example, Susie Cooper's 'Crayon Lines'. Cooper's approach reflected the belief of most pottery manufacturers that appropriate decoration was a prerequisite for good design; she believed that decoration and form must be integral: 'form, decoration and even texture in the Susie Cooper ware are part of a considered scheme; it is not merely a case of sticking a decoration on to a pot regardless of context'.[54]

At the Susie Cooper Pottery, established in 1929, Cooper produced patterns based on hand-painted dots, dashes and wavy and concentric lines. Most were produced in a single enamel colour or two colour combinations on cream-coloured earthenware. In the 1930s she developed new pottery shapes, such as Kestrel, Curlew and Falcon. These streamlined outlines reminiscent of bird forms clearly related to the undecorated forms found in modernist-inspired architecture and design, but they also revealed Cooper's knowledge of American design, gained through her awareness of that all-important export market.

In the mid-1930s Cooper cut an unusual figure in the Stoke-on-Trent pottery industry. Not only was she one of only a handful of women to own a company, she was still a young woman in her early thirties, and had already attracted considerable critical acclaim within the trade and from modernist critics alike for a number of very successful designs. She was a participant, like Nash in the SIA and she exhibited at the *British Industrial*

Art in Relation to the Home and *British Art in Industry* exhibitions of 1933 and 1935. She was a symbol of women's penetration of the design profession, but she was also emblematic of the feminization of culture in interwar Britain. In her refusal to reject the decorative, she offered a response to modernity that was quite different from European modernist exponents.

Design and Modernism(s)

Individual designers represented just one of the ways in which design responded to the new conditions of modern life. Government-initiated schemes, retailing organizations, publishing companies and commercial art/graphic design studios offered other contexts. Following several coal disputes in the early 1920s, the Sankey Commission, established by Lloyd George's Coalition Government to investigate the coal industry, recommended a reduction in working hours, a wage increase and state ownership of the mines (none of which was implemented). Less contentiously, it proposed the formation of a Miners' Welfare Fund to 'provide the miner and his family with fuller opportunities for recreation both of body and mind, with a brighter social life, and generally with a healthier and sweeter environment than the nature of his occupation can otherwise offer to him'.[55] The fund was financed by a levy of 1d a ton on coal produced and it initially aimed to provide amenities for miners, including pithead and swimming baths, recreation grounds, institutes, convalescent homes, aged miners' homes, libraries, allotments and educational opportunities for miners and their children. Socially reformist in orientation, it was reinforced by the Samuel Commission of 1926, which raised a levy of 1s. in the £1 on all mining royalties. This provided funds for a massive programme of pithead bath design and construction, leading to 345 being built between 1928 and 1939 across the coalfields of South Wales, Scotland, Kent, Yorkshire, the Midlands and Nottinghamshire, the north-east of England and Lancashire. A Miners' Welfare Architects' Department was formed in response; headed by J. H. Forshaw, it recruited young architects (male and female) at the start of their careers.[56] There were few guidelines relating to style or approach, but J. A. Dempster, head of the Northern regional office, advised them to 'Go Dudok'.[57] The work of the Dutch architect Willem Dudok, particularly Hilversum Town Hall (1928–31), used an abstract architectural language that was based on vertical and horizontal volumes and flat roofs; brick was used, with limited decoration. It proved remarkably popular in Britain and helped the Miners' Welfare architects to find an appropriate visual language for their designs, which used flat roofs, asymmetric plans and elevations and rationally planned interior designs. An attention to detail was apparent

Pithead Baths, Cardowan Colliery, Lanarkshire, designed by J. A. Dempster, 1934.

Decorative ventilation grille at Sherwood Colliery, Nottingham, designed by A. J. Saise, 1934. Such decorative detail was a hallmark of the Miners' Welfare Commission Architects' department.

in these designs, as at Cardowan Colliery in Lanarkshire by Dempster (1934) and Sherwood Colliery swimming pool by A. J. Saise (1934), which both incorporated figurative elements (such as the ventilation grille), decorative brickwork (as can be seen on the tower of the Cardowan scheme) and planting, thus ensuring that they blended with the brick-built housing typical of mining communities. Built in the industrial heartlands of Britain, these designs brought elements of modernism into the regions in a way that was paralleled with the development of multiple stores such as Marks and Spencer.

Designing Modern Britain

With shop fronts based on standardized design elements (similar to Wells Coates's Cresta shops) and modern interior planning and layout, Marks and Spencer pioneered an essentially modernist approach to design that owed as much to the USA as to Europe. Visiting the USA in 1924 to learn about retailing techniques, Simon Marks had returned to England with a raft of new ideas. He wrote: 'After my first visit, I made it my business to visit the United States as often as I could . . . It was there that I learned many new things . . . learned the value of more imposing, commodious premises'.[58] By 1939 Marks and Spencer had 234 stores on Britain's high streets, and during the economically difficult late 1920s and '30s the company had opened or rebuilt 218 stores. To facilitate this rapid expansion, company designers had developed a gold and green fascia that was abstracted and angular to suit the variables of each location. The fascias and the ground-floor window displays were mass-produced, standardized elements found in all the shops around the country. The interiors had island counters so that shoppers could examine goods easily, and lighting was modern, bright and hung low for better display. By streamlining the range of goods on sale and improving quality, the company matched the modernity of its interior design with that of its retailing policy. 'Nothing over five shillings' became the byword; through its use of synthetic fibres and direct merchandising

Interior of Marks & Spencer store on Northumberland Street, Newcastle upon Tyne, 1936. This shows the island counters and improved lighting.

strategies, Marks and Spencer responded to the mass demand for good but inexpensive clothing. Where goods were available at affordable prices, the firm attempted to improve quality; where goods existed that were more expensive than their 5s. maximum, it worked with manufacturers to bring prices down, frequently by placing large orders. As Rees put it,

> popular needs and tastes, and particularly those of the working class, were changing at a speed which we now recognize to be one of the characteristic features of the twentieth century. Any retail organisation which could interpret the public's changing needs, adapt itself rapidly to them, and satisfy them at a price within the income of the working class household was certain to receive rich reward.[59]

European modernist practice merged with traditional themes in book design and typography during this period, perhaps most notably with the formation of Penguin Books in the mid-1930s and designs for Penguin from the late 1940s by Jan Tschichold and Hans Schmoller.[60] Allen Lane published the first ten Penguin books in 1935, aiming to reprint quality fiction and non-fiction at 6d (at that time equivalent to the price of ten cigarettes), and to sell books widely, not just in bookshops, but also in Woolworths, Boots and street-corner tobacconists. Contrary to current practice, Lane wanted a simple, non-pictorial cover design, and a company employee, Edward Young, developed the initial format. The cover was divided into three horizontal sections, with solid colour at the top and bottom and a white central section for the title and author's name. The geometric simplicity of the design was reinforced by the use of Gill Sans typeface for the covers and by colour coding for book types: green for mystery and crime, orange for fiction, dark blue for biography, red for plays, cerise for travel and yellow for miscellaneous. The company identity as articulated in this design was modern, dignified and restrained, but the sense of order, simplicity and rationalism was clearly in accordance with modernist principles. This was consolidated by the appointment in 1947 of the Swiss designer Jan Tschichold, an early exponent of the new typography. Tschichold refined and standardized the basic design and symbols, and established principles of typography for designers and printers working at Penguin. Two years later Hans Schmoller took over from Tschichold, when the latter returned to Switzerland. A German citizen, Schmoller had worked in Basutoland in Africa during the war, but worked at the Curwen Press after becoming a British subject in 1946. He had corresponded with Curwen's chief typographer, Oliver Simon, while in Africa, and according to Robin Kinross, his African printing was English in orientation.[61] Schmoller had been interested in English culture and design idioms from the

An early Penguin book, *Murder by Burial* (1943). Overall design with horizontal tripartite sections first designed by Edward Young and with variations of the Gill Sans typeface for cover and spine.

The Earth Beneath Us (1958), a Pelican book incorporating Jan Tschichold and Hans Schmoller's re-working of standard 'Penguin' elements. Illustration by Victor Reinganum.

late 1930s, and from the mid-1950s he commissioned a series of wood-engravings from Imre Reiner. These formed the basis for the black-and-white vignette illustrations on Penguin covers, which referenced both English and German book-making traditions.

In direct contrast was Otto and Marie Neurath's Isotype Institute, established in Oxford in 1942 to develop visual forms that were pictorial, but also simplified and standardized. Combined with the use of the Futura sans serif typeface, Isotypes attempted to provide a universal graphic design vocabulary that was particularly effective at representing quantitative information. Isotypes were used effectively in government and related publications during and just after the Second World War – they were found, for example, in 'The New Democracy' series produced by Adprint Ltd. A title in this series, *Women and Work* by Gertrude Williams of 1945, was a good example of the Isotype Institute's modernist approach. 13 pictorial charts explored the roles that women might play in employment when the war was over. Symbols were used comparatively to show men's and women's occupations and professions, women's progress in these areas since 1911,

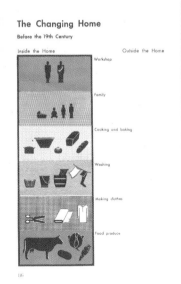

The Changing Home

Before the 19th Century

Inside the Home   Outside the Home

19th Century

Inside the Home   Outside the Home

CHART IX

Designs from the Isotype
Institute for Gertrude
Williams's *Women and
Work* (1945).

and the changing nature of the home. The book's author recommended that the reader should pay particular attention to the Isotypes, which 'are not introduced for decoration', but instead: 'if you look at them with attention you will find that they suggest all sorts of relationships between different bits of our complex society . . . It is often easier and quicker to see an argument in a picture than in words'.[62]

Isotypes, Penguin books, ceramics, pithead baths and multiple stores were all indicative of how modernist ideas permeated design across Britain, but these were strikingly diverse, and a concern for 'Englishness' – variously interpreted – coexisted and synthesized with modernist practices derived from Europe and the USA. This was underscored by a number of government initiatives by the end of the Second World War, such as the Utility schemes and the Council of Industrial Design (COID), formed in 1944. Here modernist principles increasingly framed questions of good taste and design.

Have You Good Taste?

Between 1930 and 1951 state planning and intervention on matters of taste typified many aspects of design. This was achieved through educational policies, exhibitions, government reports and surveys, collaborative projects between manufacturers, retailers and designers, and books, magazines and journals. A defining concern was the question of 'quality'. Frequently subsumed under the heading of 'taste', the focus was on design standards and public education in the wider context of anxiety about the economy, its vulnerability to foreign competition and the increasing importance of the domestic market. As C. L. Mowat put it, 'two things stand out in the economy of the '30s: increasing consumption and the development of the home market and consumer and service industries'.[63] Working-class and middle-class expenditure rose before the Second World War and resumed after 1952, but consumption patterns differed according to region, corresponding to the proportion of middle class to working class. In 1934, for example, 37.7 per cent of middle-class families were in the south-east of England and only 3.8 per cent in Northumberland and Durham.[64] From the 1930s to the early 1950s government interventions in questions of design burgeoned (particularly the Utility schemes of the war years). They included the report of Lord Gorell, who in 1932 chaired a committee examining 'the production and exhibition of well-designed articles of everyday use' on behalf of the Board of Trade.

The Gorell report was typical of attempts to 'manage' public taste and stimulate the development of good design, albeit for largely altruistic reasons. The committee comprised influential writers, concerned manufacturers, designers and critics, such as Roger Fry, Margaret Bulley, A. E. Gray, Howard Robertson and Harry Trethowan. Its aims were twofold: to examine the viability of establishing a permanent exhibition in London and organizing temporary travelling exhibitions at home and abroad, and the formation of a coordinating body to achieve this and related activities. The underlying problem was 'how best to raise the level of Industrial Art in the United Kingdom'.[65] As the report explained,

> while the enforcement of a high standard by the central controlling body should do something to induce manufacturers to produce better articles, experience indicated that such influence is unlikely in present conditions to attain its object unless powerfully supported by other and more positive measures to improve the quality of design and workmanship, and to foster an intelligent appreciation of design by the public.[66]

The committee's activities were constrained by the economic conditions of the early 1930s, although it believed that

> This is, in our view, the psychological moment, while world trade remains so depressed, for making a special effort to improve Industrial Art. Educative propaganda will, we believe, fall on more receptive ground in these times of adversity than in times of plenty;

and, at a period of relatively slack trade, time can profitably be occupied in careful planning and preparation for the future.[67]

The most significant achievement of the Gorell report was the establishment of the Council for Art and Industry in 1933, which took on an important role, producing a number of reports, including *Design and the Designer in Industry* (1937) and *The Working Class Home: Its Furnishing and Equipment* (1937). It was a precursor to the Council of Industrial Design.

 Integral to this process was the organization of a number of exhibitions by these two bodies, either directly or indirectly. Most were staged in London, although some toured Britain. Notable were *British Industrial Art in Relation to the Home* at Dorland Hall (1933), *British Art in Industry* (1935) at the Royal Academy and *Britain Can Make It* (1946) at the Victoria and Albert Museum, all in London; *Enterprise Scotland* in Edinburgh (1947); and the Festival of Britain on London's South Bank in 1951.

British Art in Industry typified these exhibitions. Organized jointly by the Royal Academy and the Royal Society of Arts (RSA), it marked a growing awareness of the importance of design, and was 'designed to show the public what an important part design plays and can still further play in the objects they habitually use and purchase'.[68] Prior to this educators, manufacturers, architects, artists and designers had debated the question of design standards in the RSA's influential magazine, the *Journal of the Royal Society of Arts*.[69] In the introduction to the souvenir catalogue, familiar arguments were outlined. It was claimed, for example, that

> with the rise of the machine, as a means to an end, there has been a corresponding fall from favour of craftsman-made goods. The main virtue accruing from machine methods is the low cost of production unknown in the days of handicraft . . .[70]

whereas crafts 'give individuality, character and charm which the machine by its very nature could not attempt to produce'.[71] Because Britain had failed to reconcile these two rival approaches, 'our markets both at home and abroad . . . have been filled up with goods of foreign competitors that have readily found buyers on account of their cheapness and of the intrinsic beauty of their conception that lies behind their design and colouring'.[72] Equally:

> 'British Made' once stood paramount throughout the world for quality and workmanship. To-day the world demands imagination as well as quality of workmanship and material. The follower of our national

pride has lately shown signs of wilting, only for lack of the fertilising effect of imagination.[73]

The front cover design of a silhouetted crown with a modern sans-serif font overlaid with flat colour blocks perfectly summed up the contradictions of national identity, modernism and modernity. The inside frontispiece with 'Wedgwood blue' background mixed quirkily hand-drawn 'crowns', the initials of the Royal Academy and Royal Society of Arts, and a glamorous colour photograph of the artist and socialite Anna Zinkeisen. Posing in a stylish interior and wearing a body-sculpting evening dress, Zinkeisen was the epitome of sophistication and glamour. The photograph entitled 'Harmony in colour' used a new full-colour process called 'Vivex' (developed by British Colour Photos Ltd) to depict an array of modern goods and services: a Cubist-style travel poster, a Moderne 'club' chair, a rectilinear occasional table and, of course, Zinkeisen herself. As though on a film set, she referenced a world of modernity. But framed by Wedgwood blue, this was quite different to Hollywood style, pointing instead to a restrained form of 'good' design and taste. The products illustrated in the catalogue represented several different approaches, but common to all was an ongoing interest in decoration, particularly colour, pattern and surface texture. This was apparent both in the furniture of Betty Joel and in numerous examples of architecture. It was evident in the hand-knotted rugs for Wilton Royal Carpet Factory Ltd and the metalwork for Mappin and

Stoneware vases and bowls designed by Vera Huggins for Doulton, Lambeth, London. Displayed at the *British Art in Industry* exhibition, 1935.

Webb, as well as in decorative and figurative glassware designs produced by Keith Murray for Stevens and Williams and the stoneware ceramics designed by Vera Huggins for Doulton.

Debates about beauty and ugliness in design permeated the literature of 1930s design in Britain. In the *British Art in Industry* catalogue, there was a full-page promotion for a new book, *The Conquest of Ugliness*, edited by John de la Valette, organizing secretary of the exhibition. With a foreword by the Prince of Wales, it included essays by crucial figures in design practice and education (Gordon Russell, Betty Joel, Alison Settle, Gordon Forsyth and Harold Curwen) in support of the exhibition and aimed at 'those who take an intelligent interest in

the everyday things which surround them'.[74] Several books were published during the 1930s on this theme, notably Margaret Bulley's *Have You Good Taste? A Guide to the Appreciation of the Lesser Arts*, published in 1933 by the same publisher, Methuen.[75] Bulley had been a member of the Gorell committee, and with Roger Fry had contributed additional memoranda in the appendix to the report *On the Production and Exhibition of Articles of Good Design and Every-Day Use*.[76] Fry, for example, recommended establishing 'Laboratories of Design', which drew to some extent on his pre-war Omega Workshops, whereas Bulley proposed a new journal and a children's school of art, the latter influenced by Paul Poiret's Atelier Martine established in pre-war France. Typically astute, Fry argued that the manufacturer had lost contact with 'educated taste', and although he was able to find and use expert advice for technical matters, when it came to the 'application of art he has no guide, no clear purpose'.[77] Taking issue with the conflation of modernism with functionalism, he noted: 'Good architecture must always remain distinct from good engineering and this principle holds equally in the design of the objects of daily use.'[78] He was, however, equally critical of 'fashion', for instance, Cubist-inspired decoration:

> You may find anywhere in our lower grade carpets and furniture fabrics a few s shaped curves and a few right angles scattered here and there across the surface for no intelligible reason and fulfilling no conceivable decorative purpose except to conciliate what is supposed to be the fashion.[79]

When it emerged in France, Cubism was 'a coherent, consistent style' that revealed 'a distinct and definite intention', but recently 'the general producer has taken a timid and side-long glance towards it'.[80] In his memorandum, Fry identified many of the problems that design reformers had highlighted since the mid-nineteenth century; partly in response, a number of practical self-help books, such as Duncan Miller's *Interior Decorating: 'How To Do It'* and Margaret Bulley's *Have You Good Taste?*, were published.[81]

Bulley's book 'seeks to make a contribution towards the training of taste in regard to the lesser arts'.[82] Good taste is determined by three main factors, 'the individual contribution, the contribution of a group or age, and the universal element'.[83] Appreciation of art could not be taught like other subjects, for the appeal of a beautiful object was directly through the mind to the eye and 'therefore cannot be put into words . . . Nevertheless something can be done by other means to free the springs of understanding and enjoyment and to create a receptive state of mind.'[84] Essentially a manual, Bulley's book synthesized established aesthetic rules and modernist ideas,

Frontispiece of the Royal Society of Arts exhibition catalogue, *British Art in Industry*, showing the artist and socialite Anna Zinkeisen, 1935.

with a peppering of qualifying observations – as with Fry – in relation to functionalism. Considering the history of furniture, she pointed out that a Stuart chair

> is not a work of art because women wore hoops and chair seats had to be wide. Neither will a modern chair survive because it was inspired by the seat of a motor-car, exhibits a new use of steel or gives no harbour to dust.[85]

Organized around comparable pairs of designs – one 'better' and one 'worse' – her argument in essence was that beauty was much more elusive than a statement of functionalism, and she firmly believed that it was possible to combine the beauty of an elaborate Queen Anne chair with the simplicity of a Le Corbusier house. She legitimized her choices and conclusions by explaining that these were subsequently endorsed by six well-known art critics or experts (Roger Fry, the directors of the Courtauld Institute of Art, the National Gallery, the Central School of Arts and Crafts and the Victoria and Albert Museum, and the editor of the *Burlington Magazine*). Adding that the purpose of the book was less about making the 'right' choices than being provoked into 'discrimination' on a subject of national importance, Bulley generally veered towards the pre-Victorian. Few examples of modern design were included and still fewer were cited as good taste.

In contrast, Duncan Miller's *Interior Decorating: 'How To Do It'* was 'a practical guide to decoration for people living in the twentieth century and using twentieth-century materials'.[86] Again using comparisons, he outlined principles of interior decoration and design that were increasingly informed by modernist discourses, particularly the insistence on designing for the twentieth century. Criticizing 'fashion', he nevertheless argued:

> Nothing would surprise the designers of the sixteenth, seventeenth and eighteenth century so much as the realisation that people were willingly submitting to the technical bonds to which they had to submit, and refusing to make use of modern materials.[87]

Comparing interiors from the same house but different periods, 1893 and 1932 (the latter designed by Wells Coates), he made clear his commitment to the 'zeitgeist'.

By 1937 the campaigning zeal of those like Bulley and Miller, combined with the activities of the Council for Art and Industry, culminated in the publication of the crucial report, *The Working Class Home: Its Furnishing and*

A double-page spread from Duncan Miller's *Interior Decorating*, 1937. Dining room and living room before and after alteration by Wells Coates.

Equipment.[88] This report summed up the preoccupation with public taste and everyday design in the 1930s and was a precursor of wartime planning and post-war initiatives. Its primary questions were:

> How far does industrial art find expression in furniture and equipment offered at prices within the reach of the working class? What proportion of the goods which their means compel the working classes to purchase are possessed of those qualities which make up good design?[89]

Making up the committee were Frank Pick, the chairman; A. E. Barnes of the High Wycombe and District Furniture Manufacturers' Federation; Elizabeth Denby, a consultant on low-rental housing; Mrs Darcy Braddell, an adviser on domestic planning; J. T. Davis, the Director of the Co-operative Wholesale Society; and A. S. Hoskin of the Board of Trade. Again the report debated definitions of 'design', and concluded that in its broadest sense design involved planning in relation to function and form. The report aimed to show how a working-class home could be furnished using well-designed products. The average working-class family income in London in 1929–30 was found to be £3 18s., but based on the assumption that wages were higher in the capital, it was decided to use £3 as the basic figure. While acknowledging that homes were furnished over a period of time, the report aimed to offer guidance in furnishing a home 'at one plunge'; thus it was accepted that the minimum requirements of a household would be expenditure of £40, although this was eventually found to be inadequate. In fact the *minimum standard* for a family of four living in a two-bedroom house with living room, small kitchen, bathroom and wc was £51 8s. 4d, whereas the *desirable standard* required an additional expenditure of £16 or £17, and to furnish a house with a parlour needed a further £30.[90] The report was in many ways a remarkable example of the 'hands-on' approach of these design reformers, and the logical extension of the activities of the Council for Art and Industry. By drawing on the expertise of those involved in design in all its stages – retailing (Davis), manufacturing (Barnes), housing (Denby) and domestic planning (Darcy Braddell), and with a skilled chair in Pick, a dedicated reformer and modernist – the report noted:

> It is possible to furnish a working class dwelling in a variety of ways with due regard to good design. It is as we thought, that good design does not necessarily enhance the cost of the article; in fact, there is a tendency in some directions for it to reduce the cost.[91]

Taste remained a perplexing issue, however. It was especially difficult to interpret public taste if popular lines exhibited 'an accumulation of patterns which is often conflicting and tiresome . . . [with a] general reluctance to look at anything bare and plain'.[92] Showing some perception of popular taste, the committee 'felt, however, that it would be unreasonable to expect the average working class home, or any other class of home, to be furnished with the uncompromising severity which some modern tastes dictate'.[93] In its conclusions the report was optimistic, since it had shown that the opportunity of living 'in pleasant, even beautiful surroundings' was not solely down to economics. Like previous exercises to improve public taste, its impact on popular taste was hard to judge, certainly during the 1930s. Instead, it was during the Second World War that the report's detailed lists of essential equipment for a working-class home proved most useful, and its guidance in matters of good taste and design potentially influential.

Design and War

Between 1941 and 1951 the overriding priorities in terms of design were supplying goods and housing to those most in need and planning the post-war economy, but matters of taste, design standards and education remained important, as demonstrated by the formation of the Council of Industrial Design in 1944 and the organization of the *Britain Can Make It* (1946) and

Designing Modern Britain

Enterprise Scotland (1947) exhibitions. Particularly important between 1943 and 1948 was the implementation of the Utility schemes, in which government controlled specific industries and raw materials so as to prioritize the supply of goods to those affected by bombing. Design standards became critical, although, as Matthew Denney has argued in respect of Utility furniture, no one approach or set of theories dominated. Indeed, it was possible to recognize both modernist and Arts and Crafts design features alongside the more typical reproduction styles, especially after regulations were loosened in 1948.[94] The designer and manufacturer Gordon Russell was made chair of the Furniture Panel in 1943 with a brief to develop a new range of designs. He had been involved in getting the first range of designs into production and in planning for further ranges; in 1946 he declared 'that to raise the whole standard of furniture for the mass of the people wasn't a bad war job'.[95] There had been a mixed reception for the first range of furniture launched in 1942; these designs were the result of the combined expertise of the advisory committee on furniture, which included manufacturers, retailers, experts on low-cost housing, the Council for Art and Industry and designers. Visually the furniture looked back to Arts and Crafts and vernacular idioms, and also referenced the popular styles of the 1930s, particularly Tudorbethan – evident in the use of dark mahogany and oak for panelling (solid wood was used for the frames and veneered hardboard for the panels). It also revealed a simplicity borne of economy (decoration was minimal, evident mainly in the handles), but also a more obvious engagement with modernity, thanks to changing tastes.

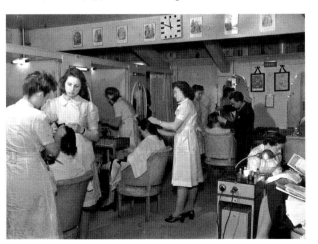

Wartime hairdresser: Steiner's Salon occupying an air-raid shelter so as to carry on business uninterrupted, early 1940s.

An early initiative of the Utility schemes, introduced in 1941, was clothing. It aimed to 'produce the nation's essential new clothing using as little power, labour and material as possible'.[96] Early designs were considered too standardized, and although in 1942 there had been attempts to raise standards of design by employing a group of well-known designers (for example, Hardy Amies, Edward Molyneux, Bianca Mosa and Digby Morton), on the whole manufacturers had not taken them up. Wartime fashion has been described as 'uniform', but the Utility schemes concentrated and designated industries in order to free up labour for essential war work. This tended to favour large companies that used mass-production techniques, such as the Co-operative Wholesale Society and Marks and Spencer, which had always depended on these processes to keep prices low.

At any rate, fashion cultures (not just clothing) conspired to subvert standards of good taste and design during the Second World War and the post-war austerity period, as women in particular glamorized their appearance. Individual wartime garments may have been boxlike with a sharp military line, but the finished 'look' was much more complex, pointing to an exaggerated femininity at a moment of intense masculinization. As Pat Kirkham argues, as the government 'exhorted all women to look as good as possible', women's magazines advised women that beauty was a duty, and make-up became widely used.[97] Styles owed much to the cinema: highly glamorous wartime images were completed by complex, elaborate hairstyles – made up of rolls, waves and cuts – topped by beautifully decorative hats. By the late 1940s and early 1950s, however, design organizations such as the Council of Industrial Design (COID), in tandem with the Board of Trade, expected women to consume, but to do so in ways that were deemed disciplined and responsible.[98] The essentially paternalistic attitudes and activities of these government bodies were nowhere more apparent than in relation to Christian Dior's New Look. Introduced into Britain some months after its launch in Paris in February 1947, the new fashion for voluminous long skirts was the antithesis of responsible consumption. It used up to 20 yards of material but required only four coupons, whereas a man's suit using three-and-a-quarter yards of material required 26 coupons (such were the inconsistencies of the residual Utility regulations).[99] It was also nostalgic, looking back to the nineteenth century, and according to some of the women MPs who entered Parliament following Labour's landslide victory in 1945, it threatened the gains made towards sexual equality, and was 'only acceptable amongst a limited class of persons and led to waste of material'.[100] Consumption was to be managed and rational; increasingly, this meant the promotion of goods that conformed to a particular design ethos – one that was essentially modernist. The vagaries of fashion generally, and the New Look in particular, were well beyond the strictures of modernist good taste and design, which not only became consolidated during this immediate post-war period, but also increasingly orthodox.

Design by Committee

Explaining the rationale for the founding of the Council of Industrial Design at its inaugural meeting on 12 January 1945, Hugh Dalton, President of the Board of Trade, cited 'a revolution of industrial design' in the USA over

Designing Modern Britain

the previous 15 years, which had 'made many of our exports old-fashioned and less acceptable'.[101] In order to give Britain an edge in an increasingly competitive world market, he argued, 'we must, therefore, make a sustained effort to improve design, and to bring industry to recognise the practical importance of this task'.[102] To make improvements in design, it was crucial to 'help industry . . . appreciate the need for good design and the training and employment of good designers', and equally 'you must encourage a discriminating home market which will give a firm basis for good exports'.[103] The economic argument was persuasive. Dalton claimed that of pre-war exports totalling £400 million, half of these were affected by design. Speaking to the members of the COID, which included Thomas Barlow (Chair), Gordon Russell, Allan Walton, Josiah Wedgwood and Kenneth Clark, he promised:

> If you succeed in your task, in a few years' time every side of our daily life will be better for your work. Every kitchen will be an easier place to work in; every home a pleasanter place to live in . . . Our export trade, and our volume of business at home, will both be the greater if our goods are planned and made, with skill and imagination, to meet the user's real need, and to give pleasure in the using.[104]

The practical outcome of this rallying call was more government intervention in design, culminating in a number of important exhibitions and reports in the late 1940s and early 1950s. Education was also widely recognized as being critical in pursuing a policy of good design, and the Council reported on 'The Training of the Industrial Designer' between February and May 1946. In an early draft, a crucial problem was identified by a manufacturer from the Midlands:

> There is a difficulty in finding industrial designers in this country who, in addition to the necessary 'flair', have a general knowledge of problems of production. There appears to be no standard of industrial designers; anybody can call himself one, and the qualification claimed may mean anything or nothing.[105]

There were many proposals in this early version of the report, which aimed at tackling first the training of designers in provincial art schools, technical colleges and regional colleges, and secondly the relationship of these to each other and to the Royal College of Art in London. But by the time the final report appeared in May 1946 there were significant omissions. Whereas the earlier version had examined questions of standards, comparability and responsibilities around Britain under a heading 'Regional Grouping of Art Schools', the final report was much watered down. A hand-written comment in the margin of the early version summed it up, asking 'how much is this our affair?'[106] It is difficult to explain this apparent shift except by noting that those in London, such as the COID, and those involved in design in the provinces were frequently at loggerheads. A particularly good example of this came with the attempt to set up a Pottery Design Centre in Stoke-on-Trent in late 1947. Characteristically, the pottery industry took great exception to outside interference by the COID in attempting to establish a design centre. As a writer in the *Staffordshire Sentinel* put it:

> If you were to suggest that a design centre would be a good thing for the pottery industry you are ipso facto telling the potters that their china and earthenware are abominable and that they don't know how to run their own businesses.[107]

The Council's activities in organizing exhibitions were perhaps more successful. Historians have discussed *Britain Can Make It* in some detail, but the ways in which the COID attempted to reach beyond the south-east of England are less well known.[108] To compensate for the fact that 65 per cent of visitors to *Britain Can Make It* came from within 25 miles of London, and to spread the message of the COID more widely (that exports were paramount, good design crucial and thoughtful consumption essential), the Council planned smaller exhibitions around Britain that aimed to link with regional or local industries.[109]

From the outset the Council had established a Scottish Committee, which planned its own exhibition when it became clear that *Britain Can Make It* would not travel north of the border. *Enterprise Scotland* was held in the Royal Scottish Museum in Edinburgh in August 1947; the architect was Basil Spence with James Gardner as chief designer. In the foreword to the catalogue, Stafford Cripps explained that the exhibition would play 'a most valuable part in the nation's export drive'.[110] The exhibition was divided into four sections, each fulfilling a specific purpose: 'Scotland Yesterday' was introductory to the whole exhibition; 'The Country' displayed sports

The 'Hall of Pinnacles' at the *Enterprise Scotland* exhibition held in the Royal Scottish Museum in 1947. Designed by James Gardner and Basil Spence.

Designing Modern Britain

goods, hotel equipment, tartans and souvenirs; 'Scotland Today' displayed commodities; and 'Scotland Tomorrow' showed plans for new towns, housing schemes, hydro-electric projects, etc. As in *Britain Can Make It*, the design was visually striking, particularly the 'Hall of Pinnacles' by Gardner and Spence displaying 'Scotland Today' commodities. Lightweight metal stands had a modern 'international' feel to them, but contrasting with this, the catalogue emphasized 'national' design qualities: pattern and intricacy, in particular, were described as 'Scottish'. Reiterating this, the exhibition included a number of traditional 'Scottish' items such as tartans and Fair Isle, although admittedly with a modern twist. Like its London predecessor, the exhibition had an educative slant on design. To reinforce this commitment to education, it was reconfigured as *Enterprise Travels*, embarking on a 1,000-mile tour of Scotland, beginning in Hamilton on 21 January 1948 and travelling to its finish in Oban on 22 May. A total of 456,000 people visited *Enterprise Scotland*, and 18,130 visited *Enterprise Travels*. Conferences were also organized and there were special events for schoolchildren, including a film entitled *A Question of Taste*.

This was a strategic moment in British design and economic development, but as Woodham argued, the COID's view of good design was based on conviction rather than evidence, and therefore unlikely to persuade the public, manufacturer or retailer.[111] These convictions were increasingly informed by modernist principles, and manufacturers in particular were sceptical about modernist aesthetics and practices. In order to overcome this, the COID organized a number of smaller exhibitions in the English regions, with a further one in Wales, as well as Design Weeks and Design Fairs in Newcastle upon Tyne, Burslem, Manchester and Cardiff. The aim

Sheffield on its Mettle exhibition, 1948, organized by the Council of Industrial Design.

was 'to arouse and maintain interest in the provinces and to supplement the Council's activities in London'.[112] To involve manufacturers directly there were links with local industries: for example, steel and cutlery in *Sheffield on its Mettle* (1948) and woollen textiles in Bradford, *Story of Wool* (1949). These continued the didactic approach of the London and Edinburgh exhibitions, with displays showing simple everyday objects of good design. A display on taste at *Sheffield on its Mettle* tried to dispel the idea that the COID was pre-occupied with only one definition of good design. Another display at the Design Fair in Manchester City Art Gallery (June 1948) aimed to show that taste varied by looking at five people and their choices of five different chairs: somewhat stereotypically, Mr Higgins the lorry driver chose a Windsor chair described as 'good and honest'. Another display forming part of the Design Fair in the National Museum of Wales in Cardiff (April 1948) showed goods costing not more than £1, purchased a few days before the opening of the exhibition, as an illustration of good design. This directly addressed criticisms of the *Britain Can Make It* exhibition, which recommended designs that were not available in the shops.

Alongside the organization of fairs, weeks and exhibitions, the COID planned Design Centres focusing on particular types of goods and materials: a Rayon Design Centre had been set up by 1948, and further ones were planned for silk, wool and carpets. By 1948 the Council felt that there was considerable evidence that greater interest was being taken in 'industrial design . . . by all classes'.[113] It was at this stage, too, that its aims were restated as Gordon Russell succeeded Thomas Barlow as Chairman. Russell brought his particular knowledge of the furniture panel of the Utility schemes to the job. He was committed to improving the quality of furniture design and his appointment at the COID was entirely consistent with this. A useful insight into Russell's views in 1947 can be seen in a children's Puffin book, *The Story of Furniture*, co-written with the Czech architect Jacques Groag. In this, it is clear that education was paramount, since they advised: 'If you are going to get good furniture when you grow up you will have to take a little trouble.'[114] They then summarized the essence of good furniture design, aiming to instil basic principles, and at the same time involve children directly: 'Will you help to show that in the new Britain nothing made by hand or by machine need be ugly, unless men and women are too careless, too stupid, or too indifferent to insist on a high standard?'[115] Although the authors argued that 'There is no reason why machine-made things should be shoddy or ugly . . . It all depends on the point of view of the people making them, the people selling them, the people buying them', Russell's Arts and Crafts philosophies inevitably spilled over:

Designing Modern Britain

It is broadly true to say that people who used at one time to make furniture by hand were interested in their product before everything, whereas many people who make it by machine are interested first in profits . . . You cannot get much pleasure out of anything which no-one took pleasure in making.[116]

The Story of Furniture was peppered with advice such as this, mixing Arts and Crafts and modernist principles. In relation to the use of metal in furniture-making, it was proposed that pleasant furniture could be made for domestic purposes from metal, but 'it must not try to look just like wood furniture'. Instead, the authors advised that metal allows 'the same beautiful precision and fit that you see in an airplane engine'.[117] Concluding with a series of illustrative comparisons (Margaret Bulley's *Have You Good Taste?* for the under-12s), children were asked, 'Isn't the simple sideboard nicer than the overdressed one?' and 'Which hall-stand do you prefer?'[118]

In the same year and written in a context of post-war reflection, John Gloag defined 'the contemporary interpretation of the English tradition' as 'exuberant and vivid' and 'changeless in character'.[119] With disregard for national unity, he focused on England, rather than Britain, reinforcing the view that the 'real English' tradition in design was the result of the enterprise and skill of gifted individuals, including Frank Pick, Gordon Russell, Wells Coates, Maxwell Fry, Keith Murray, Dick Russell, Marion Pepler and Paul Nash. Their designs in steel, plywood, aluminium, plastics, glass and textile formed one of 'the threads of the English tradition run[ning] back to medieval England, back to the wisdom of men who worked with simple tools, few materials and abundant ingenuity'.[120] Articulating a modernist preoccupation with the 'zeitgeist' or the spirit of the age, Gloag believed that by the end of the 1920s design had begun to be understood as 'industrial'.

Gloag's account rehearsed a modernist history of design first delineated in England by Nikolaus Pevsner in *Pioneers of Modern Design* (1936), except that his 'story' was peopled with British designers and architects. The antecedents of an 'English' tradition were to be found in the preceding 600 years.[121] The golden age of design, defined by him from the late seventeenth century to the early nineteenth, saw the bringing together in a 'coherent relationship the form of everything that was made, through the universal comprehension and use of rules of proportion'.[122] During this period there was no 'muddling of proportions and ornamentation', but instead 'graciousness of form while preserving that basic English characteristic, common sense, which demanded stability and delighted in good workmanship'.[123] In the intervening years, between 1830 and 1930, there had been a number of deviations from these essential rules, notably Art Nouveau, but a character-

Now, you see, we welcome the sun and we have come to regard furniture as part of the equipment for a pleasant life rather than as something to impress our friends. And we don't have more than is necessary, so that there is more room to move about. We have discovered that people are more important than things.

29

The Story of Furniture, children's book co-written by Gordon Russell and Jacques Groag, 1947. The 'preferred' modern interior.

This heavy room with its dark reds and greens and its pattern everywhere is typical of many rooms in Victorian times. Often the sun was kept out by lace curtains because it was said to fade the carpets and furniture: the effect on human beings wasn't considered nearly so important.

The 'overdressed' interior of *The Story of Furniture* by Russell and Groag.

istic of the twentieth century, he argued, was the restatement of these rules in industrial production. Citing the locomotives of the Great Western Railway and the hulls and superstructures of the Wallasey ferry boats by way of ancestry, he proposed (in a manner not dissimilar to Le Corbusier's references to motor cars and aeroplanes in *Vers une architecture*) that today the contemporary expression of the 'English' tradition could be discerned, for example, in the rolling stock, posters, stations and equipment for the London Passenger Transport Board. Acknowledging that the 'English' tradition had been 'masked by a false "Olde England"', it was 'alight and alive today all about us', and could be found in glass and steel bus shelters, radio sets, prefabricated homes, but also in Keith Murray's decorative glass, in the textile designs of Nash and Pepler, and the ceramic designs of Milner Gray and Eric Ravilious. In these, 'the spirit of England resides: exuberant and vivid as ever; different in execution but changeless in character'.[124]

Gloag's exegesis of identity and design hinted at the complexities of modern design in mid-twentieth century Britain. There was an insistent longing for an idealized 'Englishness' rooted in the countryside and dependent on traditional design values, but nevertheless dependent on new technologies (the development of crafts and the design of the interwar suburban house were examples); there was a continual interest in eighteenth-century design reworked for a contemporary market (Josiah Wedgwood being an exemplar); and alongside this were the market-driven design practices stimulated by US example (Marks and Spencer provided a case study). In addition, popular decorative design idioms were applied to a plethora of mass-produced goods that engaged with notions of modernity; and modernist theories were systematically disseminated by public and private institutions, organizations and individuals. Parallel and interwoven were debates about 'good' design and taste, abstraction and figuration in design, internationalism and nationalism, which were stimulated to some degree by the impact of Continental modernism on British, but not English, design. Design thus became a tool of economic recovery in the 1930s and '40s, while modern design practices and theories spread beyond London. The moulding of public taste was to become increasingly significant as the period of austerity gave way to economic stability and expansion, and the consumer had more disposable income, more goods from which to choose and greater opportunities to consume.

4

Designing the 'Detergent Age': Design in the 1950s and '60s

In 1951 the official guide to the Festival of Britain outlined 'one continuous, interwoven story' of Britain, and suggested that 'the people, endowed with not one single characteristic that is peculiar to themselves, nevertheless, when taken together, could not be mistaken for any other nation in the world'.[1] This chapter considers whether being modern eroded these apparent constituents of 'Britishness'. Heterogeneity and complexity characterized design in Britain in the twentieth century, and although it was 'Englishness', not 'Britishness', that mainly preoccupied manufacturers, retailers, designers and consumers, there was also an overriding desire to engage with the processes of modernity. Imagining a secure national identity was unquestionably appealing as Britain negotiated a new world role, aiming to be forward-looking and progressive. Inextricably linked to a global economy that was dominated by US economic power, the British economy had begun to turn the corner from austerity to growth following $2.7 million of Marshall Aid, and in July 1957 the new prime minister, Harold Macmillan, told a Conservative Party rally that most Britons 'had never had it so good'.[2]

In most areas of design, the development of new manufacturing, management and retailing practices meant that the consumer was the focus of the drive to sell new goods. With low levels of unemployment and inflation, Britain's economy grew from the mid-1950s to the early 1970s. More people could afford to purchase their own homes and the consumer goods to go inside. Those from the working class were now buying products – televisions, washing machines and refrigerators – that before the war had been available only to more affluent classes. New groups of consumers also emerged, particularly women and teenagers eager to buy not just domestic products, but also personal ones, such as clothes, records and magazines.

The Mini, designed by
Alec Issigonis, 1959.

SOUTH BANK EXHIBITION

LONDON

1951

FESTIVAL OF BRITAIN

GUIDE PRICE 2/6

South Bank exhibition, Festival of Britain, 1951. Sculpture in copper by Lynn Chadwick in the courtyard of the Regatta Restaurant. Chairs by Ernest Race.

Like its pre-war manifestations, modernism after 1945 was not a single coherent set of theories or practices but a number of parallel, but intersecting approaches. America as a centre of modernity was a focus of attention, but European and Scandinavian interpretations of modernism provided crucial new perspectives. Scandinavian design reconfigured modernist idioms, combining them with indigenous crafts in a manner that was very appealing to those in Britain searching for modernist languages that took account of natural materials and craft traditions. In contrast, Italian design, visually sophisticated and technologically innovative, appealed in particular to a younger market. Formally adventurous and not as obviously populist as American product design, both were admired by highbrow consumers and the design intelligentsia. Young British designers of glass, textiles, ceramics, furniture and lighting responded positively to the new approaches to design seen at the international design fairs in Milan and Stockholm. With the adoption of 'organic' and curvilinear forms, modernist design in 1950s Britain was reshaped. This reflected not only a response to Scandinavia and Italy, but also a growing awareness of the work of American designers such as Eva Zeisel, Russell Wright, and Charles and Ray Eames.

These factors influenced the Festival of Britain of 1951, where a lighter, more colourful aesthetic was first seen, but it was also the product of the socially reformist agenda dominating post-war Britain. The embryonic Welfare State provided opportunities for those who were fired by the belief that design could help to construct a better world, and in response many adopted modernist principles. This was exemplified in housing design and the development of new towns, such as Peterlee in County Durham, one of the first in the north-east of England and designed from 1948 by the pre-war modernist architect Berthold Lubetkin. He was succeeded in the mid-1950s by the painter Victor Pasmore, head of fine art at Kings College in Newcastle. But there was also an emerging critique of this utopian modernist approach, focused on the post-war Institute of Contemporary Art in London and in particular the activities of the Independent Group. The writer, critic and architectural historian Reyner Banham was particularly important in articulating this shift in attitudes, since he aligned himself with those looking beyond 1920s European modernism. He was interested in the 'second machine age', not the first, even though he was writing comprehensively about the latter in articles in the *Architectural Review* and

South Bank exhibition catalogue, Festival of Britain, 1951, design by Abram Games.

Architects' Journal throughout the 1950s. As Banham explained in his seminal *Theory and Design in the First Machine Age* (1960), 'this book was conceived and written in the late years of the 1950s, an epoch that has variously been called the Jet Age, the Detergent Decade, the Second Industrial Revolution'.[3] Observing that there was 'more than a quantitative difference between the two ages', he believed that 'highly developed mass production methods have distributed electronic devices ... over a large part of society'.[4] Unlike the technologies associated with the First Machine Age, these were not just available to an elite, upper-middle class. Instead of looking to Europe for inspiration, Banham turned west towards the USA. American consumer goods fascinated him:

> Even the man who does not possess an electric razor is likely – in the Westernised world at least – to dispense some previously inconceivable product, such as an aerosol shaving cream, from an equally unprecedented pressurised container, and accept with equanimity the fact that he can afford to throw away, regularly, cutting-edges that previous generations would have nursed for years.[5]

The American home was at the apex of post-war consumer cultures, and in Britain after the mid-1950s it was crucial in the formulation and representation of a range of ideologies regarding gender, domesticity and the family, as well as a location for the consumption of new goods (essential for the economy). In post-1945 Britain women had a dual responsibility to consume and to reproduce. Thus the *Daily Mail Book of Britain's Post-War Homes*, published in 1944 in anticipation of post-war renewal, insisted: 'housing is a woman's business. She has to make a home of the houses men build.'[6]

American mass culture – Hollywood film, literature, advertising, television, automobiles and everyday domestic goods – attracted the attention of the Independent Group, which included avant-garde artists, architects, designers and critics. Banham was a member, along with artists Richard Hamilton and Eduardo Paolozzi, photographer Nigel Henderson, architects Alison and Peter Smithson, and writer Toni del Renzio. Mass consumption and mass culture appealed to them as a way of subverting the value systems of 'high modernism' and what was seen as 'the conspiracy of good taste' promoted by the Council of Industrial Design. Elitist through their avant-garde stance, they nevertheless explored the richness and diversity of American popular culture through their art practice, writings and exhibitions. Of the last, two London shows were particularly important: *Modern Art in the United States*, which opened at the Tate Art Gallery in January 1956, and *This is Tomorrow*, opening in August at the Whitechapel

Art Gallery. The first showed the work of American Abstract Expressionists, while the latter identified two dominant tendencies in British art: Constructionism and Pop. Abstract Expressionism's 'moment' in Britain was apparently over almost as soon as it began, but it validated the existing practices of an older generation of British artists, in addition to influencing younger artists still at art school; the latter picked up on its 'American' qualities, such as scale and finish.[7] *This is Tomorrow* showed the work of artists such as Victor Pasmore, Kenneth Martin and Adrian Heath, whose aesthetic approach derived from 1930s Britain and Bauhaus-influenced ideas about the importance of the integration of art and architecture.[8] Showing alongside these were artists associated with the Independent Group, who were obsessed with developing 'an alternative modernism that engaged with the visual realities and semiotics of contemporary culture'.[9] Inevitably, there were overlaps between architects, artists, designers and craftspeople, and between different groups and theories. Artists associated with the Independent Group, Paolozzi and Hamilton, for example, taught at the Central School of Art and Design (hereafter the Central School) in London during the 1950s, before working elsewhere in Britain; Hamilton, for example, taught at Newcastle University.

As the Independent Group eulogized about American popular culture, the studio potter Bernard Leach toured the USA with his collaborators, Shoji Hamada and Soetsu Yanagi, lecturing at influential art and craft schools – Haystack Mountain School of Craft in Maine and Black Mountain College in North Carolina. They contributed to the development of an aesthetic that, in celebrating indigenous craft techniques and traditions, marked an alternative response to both modernism and mass cultures in the USA. By the 1960s, however, some perceived Leach's ceramics to be part of a tradition that looked backwards rather than forwards, and in this context studio pottery in Britain was increasingly open to new influences from US West Coast potters such as Peter Voulkos, the work of young British artists, European folk art traditions and the ceramics of Picasso. The British-born ceramicist Ruth Duckworth, taking a sculptural, hand-built approach to ceramics, moved to the USA to work at the University of Chicago. Duckworth, trained at the Central School, was typical of a new generation of potters in Britain, including Gordon Baldwin, Gillian Lowndes and Ian Auld, who had come through the art schools, particularly the Central School. The Central School produced potters who rejected the dominant oriental-inspired, hand-thrown approach and learnt instead from the Bauhaus-influenced 'Basic Design' method, which 'set out to educate, but not to train'.[10] Craft took on new and different meanings in the post-war years, and the establishment of the Crafts Centre in 1946 helped to democratize them. Becoming less intellectual, more urban and visually sophisticated (not so brown and rustic), they provided objects for different groups of consumers, thus attracting a 'new audience . . . and above all, new kinds of patronage'.[11] As in many aspects of design, the role of central and local government was critical in promoting craft, but equally social, economic and political changes were essential in providing an affluent, educated consumer for craft goods.

During the 1960s there was 'a coming into being of new subjects of history – women, gays, the third world, the working classes and youth', and proliferating design cultures helped to define these.[12] Designers, retailers and consumers were drawn from across the social spectrum, and new boutiques such as Bazaar appealed to the aspirant young who benefited from the affluent 1960s economy. British youth culture of this decade still drew on American precedents, but it became distinctive as fashion brought 'the skills of style into the world of pop'.[13] Fashion played a pivotal role in 1960s youth cultures, and unlike those of the 1950s in Britain, it targeted young women as well as men; among the iconic images of youth from this period were Jean Shrimpton, Twiggy and Michael Caine.

Growing Up in Affluence

As Liz Heron put it in *Truth, Dare or Promise: Girls Growing Up in the 50s*, 'the post-war vision of prosperity and limitless possibilities deeply underlay our everyday view of how things would be.'[14] Growing up as an only child in Scotland in the 1950s, her parents had sunk their savings into buying 'a room and kitchen with scullery', since they proved ineligible for a new council house. With no bathroom, but an inside toilet, it was also in a better area.[15] Disputing the nostalgia for the tenement streets from which they had moved, she wrote:

> This nostalgia for the Edenic community of poverty is not the nostalgia
> of working-class people but of others. Implicit in it is a belief in the
> nobility of the suffering victim and a judgement that sees the working
> class as inevitably corrupted by material things.[16]

There was considerable debate in the 1950s, as there had been in the 1930s, about the detrimental effects of mass culture and the impact of consumer goods, but as Elizabeth Roberts found in her study of the north-west of England during the 1950s, new domestic products had the potential to transform housework and women's lives. Use could be patchy and higher standards of cleanliness expected, but 'by 1950 all [her respondents] . . . had

Designing Modern Britain

an electricity supply, and domestic appliances came in a certain order. The electric iron, vacuum cleaner and television were usual by 1960.'[17]

By the mid-1950s the British economy had entered a 'golden age'. Rowntree's *Poverty and the State* of 1951 proposed that only 1.6 per cent of the total population earned less than the £5 that defined the poverty line. In addition, the Welfare State offered tangible and psychological security, particularly via house building. Winston Churchill's Conservative Government elected in 1951 – with Harold Macmillan heading the Ministry of Housing – built around 300,000 houses every year between 1953 and 1957. With incentives to builders and subsidies to local authorities, 30 per cent of new homes were private. Following the election of the Conservatives in 1955, the new prime minister, Anthony Eden, talked of a 'property-owning democracy'.[18] But, as historians have noted, the economy was 'based on shaky economic foundations', and by the 1960s there was an increasing awareness that it was not growing as fast as economic rivals, particularly Germany and Japan.[19] A new Labour government was elected in 1964 with Harold Wilson at its head, and throughout the remaining years of the decade economic problems dominated, particularly over the value of the pound. This period of economic uncertainty was paralleled with anxieties over Britain's political status, its position in the Empire and Commonwealth and its world standing, exemplified by events in Suez, Kenya, Rhodesia and Cyprus in the 1950s:

> The Suez fiasco showed that the international status Britain rightly held at the end of war was a mark of past achievement not of future potential. There were clear signs of the crumbling empire, and not less than 12 colonies, mainly in Africa, became independent in the early 1960s.[20]

Significantly, this preoccupation with the Empire and Commonwealth meant that Britain had little political engagement with Europe at the very moment that the European Common Market was set up in the late 1950s. There were economic repercussions from this, since Britain's commitment to overseas interests – particularly the Commonwealth – inhibited economic growth at home.[21]

Society in Britain between 1951 and 1967 seemed fundamentally different from that of the 1930s and '40s: 'class differences had softened and class conflict diminished. "Affluence" had lifted up the whole social pyramid, the pyramid itself was marginally a flatter and more permeable structure.'[22] While recognizing that social conflict was fragmented, there were areas of conflict, which exemplified the period, for example, between young and old, newly-arrived immigrants (black and white) and existing communities, men and women, and North and South. Caused in part by a serious labour shortage following the war, white European workers, black British citizens and Irish immigrants came to Britain in large numbers to work in coal mining, agriculture, textiles, domestic service and transport. Significantly, those 'from countries such as Poland, the Ukraine and Italy were aliens and therefore had few rights in British society . . . in contrast black workers from the Commonwealth had British citizenship', and even in a peak year such as 1956, just 30,000 people from the West Indies arrived in Britain, in contrast to 60,000 from Ireland, but particular areas of the country experienced changing patterns of settlement.[23] Generally moving inland from ports into industrial areas in the North and the Midlands, as well as specific districts of London – Southall, Lambeth and North Kensington – black immigrants, in particular, experienced racism and discrimination that permeated most aspects of life in Britain. Two Commonwealth Immigration Acts were passed in 1962 and 1968 respectively, the first by a Conservative government and the second by Labour; undeniably racist, these set the tenor for post-war Britain. Concern about black settlers in the late 1940s and '50s clustered 'around a distinct range of anxieties and images in which issues of sexuality and miscegenation were often uppermost'.[24] In spite of this, relationships were forged within new and existing communities, and recent immigrants contributed energy and dynamism, as well as the richness of different cultures, to building post-war Britain. Gail Lewis described her upbringing in 1950s Britain: the daughter of a Jamaican father and a north London, white mother (with a Scottish grandmother), she was brought up to eat porridge (with salt), pie and mash, and saltfish and ackee.[25] Gloria Bennett, supplementing her job as a bus conductor with income from home dressmaking, settled in Doncaster in South Yorkshire when she arrived from Jamaica in 1959. There she produced fashion designs for the black female community, bringing to British dress codes 'idiosyncratic inflections' representative of 'Jamaicaness'.[26] Bennett was noted for her skill in designing for older women, and perhaps a crucial factor of the period was the increasing generational divides, particularly between young and old, of which fashion and dress were important indicators.

The emergence of a distinct group, 'youth', has been attributed to this period, 1951–67, linked to increasing affluence and a greater propensity to consume, and the emergence of youth cultures such as the Teddy Boys in the 1950s certainly supports such a view. These young men were the epitome of 'streetwise'. Between the school-leaving age of 15 and conscription at 18, a window of opportunity opened, particularly for those who, in full-time work, had money to spend prior to marriage and family responsibilities.

Designing Modern Britain

New films and radio stations, pop music and clothes enabled them to articulate a different, more modern identity from that of their parents and grandparents. Inspired by American film and black R&B music, but combining elements from upper-class Edwardian menswear, young men consumed all manner of new goods, but particularly clothes and records. They constituted but one of a number of new markets emerging in this period, taught to consume via advertising, television, pirate radio and magazines. Women were another.

The question of appropriate gender roles for men and women was debated in a variety of contexts during this period, and inevitably these were inflected by parallel discussions about race and class. Women's place in society attracted a good deal of attention, especially following the disruption to gender roles during wartime. One of the titles in The New Democracy series – *Women and Work* (1945) – posed questions about what part women would play in the post-war world.[27] Its author, Gertrude Williams, explained: 'The war had made an immense difference to women's lives, both with regard to the kinds of work they do and, perhaps even more, with regard to public opinion concerning them.'[28] She acknowledged that the place of women had been taken for granted before the war, and that the war had challenged this, but warned against running from one extreme to another by proposing that there were no differences between men and women:

> Men and women are different because of the different parts they have to play in producing the next generation, and we have to keep this in mind when considering the role that women do, can, or should fill in the world of employment.[29]

Herein lay the dilemma for women. How to have full opportunities in life without these being curtailed by patriarchal assumptions about women's role in reproducing the family?

In 1951, 36 per cent of women were in paid work outside the home and by 1961 this had risen to 42 per cent, but most telling was the sharp increase in married women in work, from 26 per cent in 1951 to 35 per cent in 1961.[30] Also, although many more women remained in the workforce after 1945 than after 1918, most were in part-time work. Government was keen to retain women in the labour force, and 'in the immediate post-war decades married women left work at the birth of their first children and returned to work when the children left school'.[31] But messages were conflicting after the war. Women wanted to work, but also to have the choice to leave work to have children and to have shorter working hours so as to care for them.

In the embryonic Welfare State women's work was therefore concentrated in part-time jobs and the service sector, doing particular types of work 'because of profoundly gendered ideas as to what kind of work is appropriate for women'.[32] Much of this low-paid work was also low status. In tandem with the pull for women to work, albeit in particular types of jobs, was a growing ideology of domesticity, evident in women's magazines and reinforced by influential writers such as John Bowlby, who highlighted the apparent problem of maternal deprivation in the children of working mothers. In the mid-1950s, 58 per cent of women read magazines, particularly *Woman* and *Woman's Own*, which included features on home, beauty and family. Sales of magazines soared, but 'enjoyment of women's magazines [did] not necessarily imply that their readers became solely domestically-oriented fashion slaves'.[33] Sexual politics were discussed in numerous contexts during the 1960s, with important legislative change coming with the Abortion Act in 1967 and the Divorce Act in 1969, but 'contrary to the popular view of the 1960s as a decade of sexual licence, marriage proved increasingly popular'.[34] New technologies of contraception such as the pill may have offered women some measure of control over reproduction, but this seemed largely theoretical, since by 1969–70 'one third of teenage brides were pregnant and 43 per cent of all births conceived premaritally were to teenagers'.[35] The 'permissive moment' in the 1960s was remarkably short and the changes were ambiguous for women. Becoming a woman after 1945 required an ongoing negotiation of conflicting desires, expectations and ambitions. Inevitably, the social and political transformations brought by the processes of modernity engaged new groups of people – women, young people and immigrants – in continuing dialogue with new design practices that were predominantly modernist in orientation.

Festive '50s?

As a visual style and as a way of thinking about design and architecture, modernist theories and practices gained in popularity, stimulated in part by the Festival of Britain. At the same time there was considerable debate over the extent to which the exhibition was modern, and whether it represented 'Englishness' and/or 'Britishness'. Twenty-five years later at an exhibition at the Victoria and Albert Museum commemorating the Festival, Reyner Banham believed that there was nothing 'English' about the style of the exhibition.[36] To him, it had referenced Italian, Scandinavian and American design – a point he made in relation to Ernest Race's Antelope Chair. For Banham, its spindly, splay-legged 'insect-look' silhouette derived from Charles Eames's DCM-1 chair, but referencing a traditional type of English

Designing Modern Britain

chair; it was 'like a drawing of an armchair done in steel rod, but still in the Eames style'.[37] Others have taken a different view: 'From the outset, the Festival planners in London were very committed to and interested in presenting a modern Festival of *Britain*, not a conservative, *volkisch* rendering of "Deep England."'[38] This particular analysis acknowledged the earnestness of the Festival, since it was 'simultaneously a public celebration, an educational undertaking, and a constructed vision of a new democratic national community'.[39] With events and activities across Britain, there was also considerable diversity. The South Bank site on the Thames was the centre of activities, but there were exhibitions elsewhere in London: *Science* at South Kensington; *Architecture* at Lansbury, a bomb-damaged site on the East India Road; a *Festival of British Films*; and the *Festival Pleasure Gardens* at Battersea. Up and down the country there were numerous events, exhibitions and displays: in Wales, a pageant and folk festival in Cardiff, and the Welsh Hillside Farming Scheme at Dolhendre; in Northern Ireland, the *Ulster Farm and Factory* exhibition, and an arts festival in Belfast; and in Scotland, exhibitions on *Industrial Power* and *Contemporary Books* in Glasgow, and *Living Traditions: Scottish Architecture and Craft* in Edinburgh. There were also arts festivals at Eisteddfod, St David's, Perth, Inverness, Aberdeen, Dumfries and a 'Gathering of the Clans' in Edinburgh. The Festival ship, *Campania*, a 16,000-ton escort carrier, took the *Sea Travelling* exhibition to Britain's ports – Newcastle, Bristol, Birkenhead, Southampton, Hull, Plymouth, Dundee, Glasgow, Cardiff and Belfast – for two-week periods, while the *Land Travelling* exhibition visited Leeds, Birmingham, Nottingham and Manchester.

In this context, the concern with 'Britain' rather than 'England' was progressive and modern, not just visually but ideologically, seeking to extend the wartime spirit of unity; it thus chimed with the new Labour government's vision of a democratic post-war Britain.[40] The staging of so many regional activities suggested a keen awareness of regionalism, albeit stereotypically depicted via their traditional industries and located within an overarching concern for national unity.[41] But equally, this attempt to recapture the 'national community' of wartime could be interpreted as indicative of disintegration and fragmentation as regional economic inequalities came to the fore yet again. The exhibition symbol and poster designed by Abram Games summed this up, referencing a resurgent Britannia donned in blue, red and white and with specific class connotations.[42] Regional identities may have been central to the representation of post-war Britain, but in Games's poster Britain is a uniformly green land with no identifiable boundaries except for the coastline, showing little concern for Europe, or indeed empire. Reinforcing this, the typography design panel used a 'British' typeface based 'on the Egyptian types cut by Figgins, Thorne and Austin between 1815 and 1825'.[43] Popularly depicted as 'a tonic to the nation', Adrian Forty observed dryly that a tonic was given to a recovering patient, whereas Britain was hardly in recovery. Indeed, the Festival concealed Britain's economic and political weaknesses.[44] But with more than 8.5 million people visiting the South Bank, most writers agreed that the show was a spectacular success, even though there has been considerable debate about its originality and influence, especially in terms of design, a central preoccupation. Banham thought it had 'influence-wise . . . died a-borning', whereas Paul Reilly, Director of the COID, believed that it marked a turning point in public taste.[45] A young critic at the time, Banham did recognize that 'the Festival showed the way if not the style' to the 1960s generation.[46]

Photographs of the exhibits, display stands, pavilions, plan and land-scape, and the typography of the exhibition, show that there were a number of common visual characteristics – new materials, bright colour, geometric patterns and abstract forms – as well as an interest in science, engineering and technology that found its way into design (visually with crystalline structures, but also structurally). It was in the pavilions, landscaping, sculpture and murals that the 'look' of the Festival was best articulated. There was visual continuity, particularly in the patterns, form and colour, and an attempt at synthesis across these, with numerous collaborations between artists, architects and designers. Jane Drew's Riverside Restaurant, for example, had a mural by Ben Nicholson; the interior of Misha Black's Regatta Restaurant had a ceramic mural by Victor Pasmore, while outside there was sculpture by Lynn Chadwick and a garden by Maria Shepherd. Barbara Hepworth's sculptures were in the Dome of Discovery and in the garden courtyard of the Thames-side Restaurant, and those of Henry Moore were sited near the Country Pavilion. Equally, the steel and aluminium Skylon by Powell and Moya, the Dome of Discovery by Ralph Tubbs, the Telecinema by Wells Coates, and the Waterloo Bridge Gate by Jane Drew and Maxwell Fry were paeans to technological development and engineering advances. Underpinning the design ethos was a distinctive political commitment to social welfare; the New Schools Pavilion by Fry and Drew and the housing at Lansbury, for example, typified a concern for design that embraced everyone – in these particular instances, a vision for a new educational democracy and a reconstructed post-war Britain.[47]

Zeal characterized the design processes of the exhibition, and further highlighted the propagandizing role of the COID. Journals such as *Design* magazine (begun in 1949) and the 'Things We See' series (published by Penguin in 1947) were vehicles for the COID to expound its principles and to support Festival planning. The pamphlet *The Things We See: Houses* summed this up. Dealing with housing design in 1947, it proposed that architects had picked up where modernist precursors in the 1930s had left off, 'direct and candid, delighting in the mechanical and contemptuous of the artificial'.[48] Addressed to the needs of the consumer – 'the man or woman who pays the piper' – the author, an Eton- and Oxford-educated architect, Lionel Brett, set about establishing guidelines for post-war housing. Light, bright, plain and clean mass-produced houses were proposed by this Georgian-house resident from rural Oxfordshire, described by Banham as part of 'the officer-and-gentleman establishment'.[49] Under COID control, commissioned architects were invariably modernist: Misha Black, Jane Drew, Maxwell Fry, F.R.S. Yorke, Wells Coates, Robin Day, Lucienne Day, Ernest Race, F.H.K. Henrion and Maria Shepherd. For many younger designers and critics, the Festival style was anathema; instead, they preferred Le Corbusier's *beton brut* or Pop Art. As Banham had put it, the 'modernity' of the exhibition was 'old hat'; 'in spite of the semi-official line that the Festival was by British Originality out of Stockholm 1930, the exhibition it most resembled was the Triennale di Milano of the same year'.[50]

The impact of these various 'modernisms' – 'Festival Style', Scandinavian and Italian – was increasingly evident as the 1950s progressed, producing a 'New Look' in design.[51] Visual sources for a more organic, curvilinear modernism included Eva Zeisel's ceramic tableware 'Town and Country' for the US company Red Wing Pottery in 1946; Tapio Wirkkala's glassware for the Finnish company Iittala in the early 1950s; Gio Ponti's designs for Ideal Standard bathrooms in the early 1950s; Harry Bertoia's Wire Chair (1952) and Eero Saarinen's 'Tulip' chair (1956), both for the US furniture manufacturer Knoll. Equally important were the close links with artists and sculptors such as Hans Arp, Constantin Brancusi, Henry Moore, Barbara Hepworth, Alexander Calder, Alberto Giacometti, Joan Miró, and Naum Gabo and his brother Anton Pevsner whose work was innovative in form, materials and technique.

By the end of the 1950s new styles were pervasive, undercutting regional and national differences as the pressure increased for designs that sold abroad. The ceramics manufacturer Midwinter was indicative of this. Like several other British pottery manufacturers, the firm relied on the North American market for exports and found adjustment to post-war economics difficult. While many in Britain had been producing Utility ranges, in the USA pottery manufacturers had developed new tablewares free of wartime restrictions. Companies such as Iroquois China, Steubenville, Castleton China, Red Wing and Hall China employed Russell Wright and Eva Zeisel to produce new shapes, colours and patterns, such as 'American Modern' (Wright for Steubenville, 1939), 'Casual' (Wright for Iroquois China, 1946),

'Museum' (Zeisel for Castleton China, 1946) and 'Tomorrow's Classic' (Zeisel for Hall China, 1952). These were promoted differently as practical 'oven-to-table' wares and packaged for different lifestyles and the various stages of people's lives – for example, 'starter sets' for the newly married. Organic curvilinear shapes were adapted to flatware (plates and saucers) and to hollow-ware (cups, bowls and serving dishes) in complementary, interchangeable bold colours. The glazes were often unusual too – matt, speckled, white or one solid colour. Aware of these new trends and developments as a result of a visit to the USA in 1952, Roy Midwinter, the company manager, realized how dull and parochial Midwinter's export range seemed. Back in Stoke-on-Trent he set to work, updating product ranges, modernizing patterns and developing new shapes for a younger market. Two new shapes, 'Stylecraft' and 'Fashion', launched in 1953 and 1955, were his response to this. To accompany these, new modern patterns were developed by the in-house designer, Jessie Tait ('Festival', 1955), and freelance designers, such as Terence Conran ('Plantlife', 1957). Midwinter was an excellent example of a mainstream British manufacturing enterprise both adapting to the changing export and economic conditions after 1945 and recognizing the importance of design in this process.

By the end of the 1940s it had become clear to many associated with the COID that a new type of industrial designer was needed: one attuned to changing markets and consumer behaviour, as much as to new technologies and mass-production systems. In response, debate about design and craft education became more focused, and inevitably the didactic tone of the COID permeated popular newspapers and magazines. Education also involved the consumer, and there were numerous exhibitions following the Festival of Britain that attempted to promote and define what constituted well-designed products, either hand- or machine-made. After the formation of the COID in 1944, a Crafts Centre had been established in 1946, joined by the Crafts Council of Great Britain in 1964, which contributed to these discussions. Initially, it was the education of designers that preoccupied practitioners, policy-makers and educators, the central questions being first how best to train designers for a role in the modern world and second what that role should be. Tied to this was the question of the aims and purpose of the Royal College of Art.[52] From 1948 its head was Robin Darwin, the author of an important report written a couple of years earlier under the auspices of the COID. The *Report on the Training of the Industrial Designer* had argued that art schools had become too detached from industry, and addressing the question of appropriate training for designers, it reiterated the Royal College's fundamental purpose – to train designers for industry, not to produce fine artists. As a consequence of this and other related

Designing Modern Britain

activities, the college was reorganized into a number of schools to enable specialization in a specific field of design: Ceramics; Textiles; Typography and Design for Publicity (subsequently Graphic Design); Silversmithing, Metalwork and Jewellery; Fashion Design; and Light Engineering and Furniture (subsequently Woods, Metals and Plastics). These changes marked a very conscious move away from the models of art and design education whose origins lay at the end of the nineteenth century, and a clear focus on the needs of industry. Several well-known designers were brought in as teachers: Richard Guyatt (graphic design), David Pye (furniture), Robert Baker (ceramics) and Madge Garland (fashion).

There were numerous overlaps between design and craft after 1945, but craft remained an important, diverse and autonomous activity. At the launch of the Crafts Centre, there had been considerable emphasis on the role of crafts in improving industrial design, and there were several practitioners, best known for industrial design, who were also committed to crafts. In the work of David Mellor and Robert Welch, for example, it was not possible to separate these activities neatly, but there were some fundamental philosophical differences within the craft world on how best to train craftsmen and women. In particular, there was a split between those who advocated workshop training in the crafts (as had been common in pre-war Britain in the workshops of, for example, Bernard Leach and Ethel Mairet)

Large plate, W. R. Midwinter Ltd, Stylecraft range with 'Primavera' pattern designed by Jessie Tait, 1954.

and those who demanded art school education in the crafts.[53] Young craftsmen and women in the 1950s rejected the approach of those such as Mairet,
preferring 'not to boil up leaves in buckets', seeing themselves instead as
designers who used hand processes.[54] Particularly important to this were
those art schools – such as the Central School – that taught 'Basic Design',
the 'antithesis of the Leachian workshop approach'.[55] Following Bauhaus
methods, students were encouraged to work in experimental ways with a
variety of materials, some of them unconventional. This led to an unortho-
dox and questioning approach to the crafts, rather than an emphasis on the
acquisition of craft skills. The Central School in particular advocated 'Basic
Design' methods with some success, and it employed a number of artists as
teachers in the 1950s, including Pasmore, Hamilton, Paolozzi and Alan
Davie – echoing 1920s Bauhaus practice when artists such as Kandinsky
and Klee had taught the preliminary course. Davie's work, for example,
spawned an interest in pre-modern pattern, decoration and form, which
influenced the work of several potters, including Gordon Baldwin and Ruth
Duckworth. Baldwin recounted that Paolozzi encouraged students to mess
with clay, and not to worry 'about ceramic quality and feeling for clay'.[56]
This approach overlapped with an interest in crafts from other cultures. In
addition to South American and archaic ceramics, French and Italian country wares were admired alongside the ceramics of artists such as Picasso,
whose work in clay had been shown in an exhibition staged by the Arts
Council in 1950. These new visual styles had strong affinities with the work
of contemporary industrial designers, pointing to a new vibrancy and energy in crafts, adding colour, pattern and sculptural form to the restrained
modernism of the post-war home.[57] New clients also commissioned crafts
as part of large prestigious public projects, such as the rebuilding of
Coventry Cathedral by Basil Spence in the early 1960s. Contemporary crafts
were featured more often in design and women's magazines and important
national collections, such as the Victoria and Albert Museum. Smaller craft
galleries were set up to provide goods for the post-war interior. Returning
from Italy after the war and inspired by the brightly-coloured Italian ceramics he had seen while abroad, in 1945 Henry Rothschild established a small
gallery and shop, Primavera, to sell unusual crafts and goods. Situated on
Sloane Street in London, it sold an eclectic mix of folk art, rural crafts, mass-
produced design and progressive crafts. Rothschild, committed to
promoting the 'best things whether hand made or machine made', was an
important patron to a number of younger potters, including Lucy Rie and
Hans Coper.[58]

Ideal Homes: Design and Equipment

In a number of ways, the home provided the space where the post-war modernist vision of a better world was best articulated and in which the integration of design – architecture, crafts, interiors, products and furniture – was achieved most effectively. A popular vision of the post-war home could be seen at the *Daily Mail Ideal Home* exhibitions. Begun in 1908, the exhibitions set out to portray the ideal home – its conception, construction and equipment – and to 'put before the housewife the products of the best brains of the building and house equipment industries'.[59] In 1944 a comprehensive survey produced the *Daily Mail Book of Britain's Post-War Homes*; based on the ideas and opinions of the 'Women of Britain', it proposed: 'unless women take more active and objective interest in housing, the small rooms, the inconvenience and lack of efficient planning will be repeated'.[60]

In response to the survey, a woman architect, Barbara Auld ARIBA, was brought in to produce 'the house that women want'.[61] The essential requirement was for a three-bedroom house, well built in a tree-planted cul-de-sac, part of a group but not one of a long repetitive row. In reach of all amenities – shops, churches, health centre, schools and clinics – the house was to be well connected to public transport. With the kitchen overlooking the front of the house and large windows from the sitting and living rooms looking over the garden, the intention was that children could be watched while the housewife worked in the home. Care was given to the layout of interiors and the positioning of services and household technologies, such as plumbing, electricity, gas, water and heating, as well as the arrangement of kitchens, utility room, bathroom and WC. The first floor had three bedrooms, two with built-in storage and linen cupboards, a bathroom and separate toilet. The ground floor had three rooms: a small sitting-room, a large living-room, with space for a dining table, and a kitchen with ancillary washhouse/utility room. There was considerable debate about whether the kitchen should open onto the living room or be separate. In the survey 64 per cent of women wanted a separate kitchen, and although the architect's plans showed the kitchen opening onto the living room, the plan was structurally arranged so that an alternative design with separate spaces would be possible. What women wanted most was space combined with careful planning: 'Thus, if the sink, draining board, cooker, and larder be the only fitments originally supplied, let the architect and builder so site them that the other standardised fitments can be efficiently installed later.'[62] There was a consensus that a living room and a sitting room were both necessary to allow for privacy and quiet for different members of the family. Both architect and author clearly held the opinion that women's desire for a separate kitchen stemmed from memories of the dark little rooms from which they wanted to escape, rather than modern kitchens that have 'modern machinery and the assistance of science'.[63] The kitchen was the epicentre of the architect's, author's and women's vision of the future family home. It was rational, planned and equipped with flat, smooth work surfaces, so that dirt and dust could be eradicated with ease. The hard work of the house was reserved for the washhouse or utility room, in which would be found the washing machine, sink, wringer and boiler, as well as a place to dry and iron clothing; accessible via this would be a solid-fuel store and a WC.

The vision of the ideal home proposed by the *Daily Mail Book of Britain's Post-War Homes* was one recommended for women across Britain. To coincide with the *Daily Mail* survey, an exhibition of *Northern Housing* was staged in Newcastle upon Tyne in December 1944. Organized by the local authorities of Northumberland, Durham and the North Riding of Yorkshire, in collaboration with various government departments, this promoted steel factory-made houses as a solution to the impending housing crisis. Housing schemes, model homes and interiors were proposed by several local Urban District Councils, including the South Bank Housing

'The House Women Have Chosen', *Daily Mail Book of Britain's Post-war Homes*, 1944.

Designing Modern Britain

Scheme at Eston in the North Riding of Yorkshire; the Longbenton Housing estate proposed by Newcastle Corporation; and a model kitchen by Stockton-on-Tees Corporation. The gas industry promoted 'The Practical Northern Home', again designed by a woman, Mary Proctor Cahill ARIBA (of the Alnwick firm, Reavell and Cahill). This particular home was aimed at 'the great masses of people who may be said to be of moderate means', and it was 'practical for north of England conditions in every sense of the word'.[64] The separate kitchen debate was considered by Cahill: 'shall the kitchen be the workshop of the home, or shall it be part-kitchen part-living room?' Her solution was flexibility. A dining recess off the kitchen with a door from this to the living room allowed some respite for the housewife from the 'isolation sometimes felt when the kitchen is entirely [a] workshop'.[65] The idea that women architects had a particular aptitude for design

'The Practical Northern Home' at the *Northern Housing* exhibition staged in Newcastle upon Tyne in 1944.

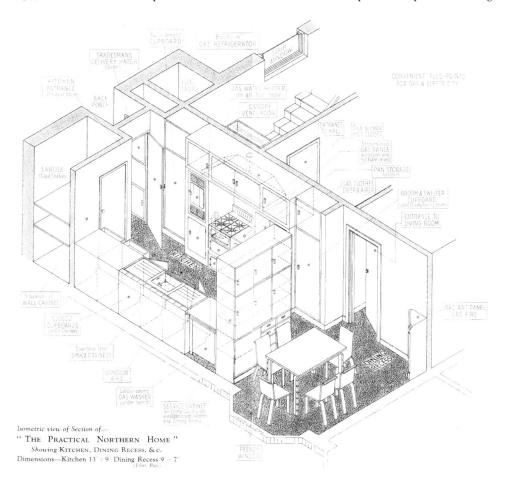

Isometric view of Section of—
" THE PRACTICAL NORTHERN HOME "
Showing KITCHEN, DINING RECESS, &c.
Dimensions—Kitchen 13´ × 9´ Dining Recess 9´ × 7´
(Plus Bay)

for women, evident in both the *Daily Mail Book of Britain's Post-War Homes* and in this leaflet, was indicative of a form of stereotyping within the profession, but this meant that after the war women were well placed to gain architectural jobs in the public sector, designing housing, schools, hospitals for local authorities, etc.[66] It was clear from the literature that it was women's role to run the home, not men's, and there was no question of its importance: 'With happier and more spacious homes the adolescent boy and girl would feature far less frequently as young criminals because they would have a real home in which to make their own fun.'[67]

European Lighting exhibition, Scottish COID, Glasgow, 1950.

Consuming Cultures: Good Taste and Pop

In the aftermath of the Festival of Britain, the COID appeared to consolidate its position as arbiter of good design in Britain. For the exhibition, it had drawn up a stock list of more than 20,000 'exhibition worthy' products of British industry that – combined with its index of industrial designers and its magazine *Design* – seemed to ensure its authority in defining standards and values in design in Britain.[68] By 1956 the COID had also opened the Design Centre in London as a permanent exhibition space for well-designed goods, followed by the opening of new premises for the Scottish Committee of the COID in 1957. Exhibitions organized by the Scottish Committee had included *European Lighting* in 1950, which showed lighting by the Milanese and Danish manufacturers Arte Luce and Le Klint, with fittings by the Danish craft gallery Haandarbejdets Fremme. These designs codified the simplified and restrained post-war modernism preferred by the COID, but by the end of the 1950s its determination to assess and prioritize what was important in design was being questioned by consumer representatives and undermined by a growing perception that it was indicative of a 'narrow middle-class taste'.[69] How widely this view of the COID was shared is unclear. Jonathan Woodham proposed that such views and criticisms probably had little impact on the average consumer, but the crucial point was that because of its semi-official position, the COID became the target of those who believed its modernist outlook outmoded.[70] This may well have been the case, but its propagandizing purposes continued through the 1960s. Nevertheless, between March and April 1969 it relished the opportunity to take *The Design Centre Comes to*

Designing Modern Britain

Newcastle exhibition to the furniture store Callers, situated in the city's premier shopping area, Northumberland Street, and the adjacent Saville Row.[71]

Founded in the early 1920s by Abe Caller, the son of an East European Jewish cabinetmaker who had set up business in Newcastle in 1897, Callers was one of a number of established furniture retailers in Newcastle (others included Chapmans and Robsons).[72] Refurbished in 1954–5, the shop frontages and showrooms in Northumberland Street and Saville Row were distinctively modern.[73] Following the death of their father in 1958, Roy and Ian Caller had taken over the day-to-day management of the family-owned company. Like their father, they bought from well-regarded manufacturers in High Wycombe, such as Ercol Furniture Ltd and E. Gomme Ltd, Harold Lebus in London and Meredew in Letchworth, but they were also 'intrigued by modern design'.[74] In particular, they liked the simplicity, modernity and quality of Scandinavian modern design, especially that from Denmark. Regularly visiting the Copenhagen Furniture Fair and sourcing the goods on display for their Newcastle shop, they also admired the interior design and products of Illums, one of Copenhagen's foremost department stores.[75] Their gallery of model rooms was an innovative means to display 'ideal' rooms that incorporated furniture, textiles, lighting, ceramics, glass and even 'art' so as to show relatively inexperienced consumers of modern design just how to make a 'modern' home.[76] Callers' plan to bring the Design Centre exhibition to Newcastle was indicative of their enthusiasm for design, but it is worth remembering that Roy and Ian Caller were experienced furniture retailers with extensive knowledge of the furniture trade.

The COID appointed the architect/designer Helen Challen to design the exhibition, and Ian Caller recalled the battle of wills as he and his brother toned down her radical ideas for a North-East market.[77] With the construction of six fully furnished rooms (including a kitchen), as well as other displays, exemplary modern design was shown at *The Design Centre Comes to Newcastle* exhibition over four weeks in the spring of 1969. Advertising the exhibition in the local paper with the slogan 'Why not a Chair of Paper or Foam', consumers were invited to come and see Frederick Scott's foam seating for Hille, David Bartlett's slot-together paper chair 'costing little over £2', and Quasar Khanh's clear plastic inflatable chair.[78] Other exhibits included Harry Bertoia's sculptural metal basket chairs, Peter Murdoch's hexagonal plastic chairs, stools and low tables for Hille, knockdown furniture for children, as well as Fidelity radios and Philips automatic washing machines. Announcing that 'what London has today Newcastle has as well', the 'Trend' section of the *Newcastle Journal* gave positive encouragement to modern design, and emphasized the importance of the Design Centre as a 'must-be-seen' place in London.[79]

Opened by Sir Paul Reilly and with an accompanying lecture to architects and designers at the Northumberland Street store by Gordon Russell, education was a dominant theme of the Design Centre exhibition – both for the consumer and designer. Alongside the Design Council exhibits were prototypes by students of the Newcastle College of Art and Industrial Design. Several of these revealed the students' knowledge of recent trends in furniture and hinted at the widening gap between the paternalistic approach of the COID and 1960s Pop culture, as was hinted at in Russell's lecture. (Upon being questioned about the apparent draughtiness of an exemplary modernist bus shelter that he had used to illustrate his lecture, Russell's own class perspective came to the surface as he replied that since this audience rarely took a bus, it wasn't a serious concern.)[80] In the advertisement for the exhibition in the *Evening Chronicle*, a number of points were emphasized that indicate the COID's priorities: first, the promotion of 'Britishness' via British-made goods; second, the importance of goods being 'designed' by designers, thus 'names' were prominent – textiles by Margot Shore and seating by Frederick Scott, Harry Bertoia, William Plunkett and Robin Day; and third, the insistence on design standards denoted by the Design Centre 'kite' symbol. Nevertheless, the throwaway and disposable chairs, as well as the vivid colours and vibrant patterns, were harbingers of the impact of Pop cultures and the needs of new markets that were to be comprehensively addressed. Resulting from a suspected electrical fault, Callers experienced a devastating fire in late 1969 that necessitated its complete rebuilding. Reopened in 1971, it had a plethora of new design features and, just as importantly, new sections,[81] including a travel agency (within the furniture store), to provide affordable 'package holidays' abroad for more cost-conscious consumers, and a record department, which brought in the youth market. The interior design of the reconstructed store was flamboyant, and specific features (brick curved walls and glass floors) were borrowed directly from the Copenhagen store Illums.[82] Open-plan with brick, steel and glass, internal water features and themed rooms, the store was an example of modern design in Newcastle with broad appeal. The stunning lighting department, again inspired by Illums, brought a visual sophistication to the increasingly difficult business of selling furniture. Ian and Roy Caller understood that the didacticism of the COID would not work. Equally, their displays showed an awareness of the disposable or futuristic aesthetic associated with the Independent Group, but were a far cry from the radical avant-gardism epitomized by it. Firmly rooted in the competitive retailing marketplace of Newcastle, Callers offered instead an array of desirable domestic products – modern and traditional – that provided essential furnishings for various 'Ideal Homes' typically comprising three bedrooms, living and dining rooms, and a kitchen. New house building had gathered pace in post-war Newcastle. In addition, several New Towns were built within the region: on Newcastle's outskirts were Cramlington, Killingworth and Washington New Town, and within travelling distance in County Durham were Newton Aycliffe and Peterlee. Not only did Callers have branches in Blaydon, Morpeth, Consett, Hebburn, South Shields and Ponteland, but also in Washington New Town and as far south as Middlesbrough, thus capturing the regional markets. As new homes needed new categories of goods as well as traditional ones, Callers was on hand to guide relatively inexperienced consumers in their domestic purchases.

By the late 1950s and early 1960s opposition to the COID was informed by the ideas of the Independent Group at the Institute of Contemporary Arts in London, among others. Alison and Peter Smithson, who had worked collaboratively since graduating from Newcastle University in the late 1940s, had contributed to *This is Tomorrow* at the Whitechapel Gallery in 1956. A celebration of American mass culture and a technological future, this exhibition – occupying a seminal position in British art – also played a pivotal role in helping to reorient design in the late 1950s and '60s. Acknowledged as the birthplace of pop, *This is Tomorrow* was just as much concerned with design, but its design trajectory was very different from that of the COID.[83] Alison and Peter Smithson's 'House of the Future' at the *Daily Mail Ideal Home* exhibition in 1956 summed up what sort of design they envisaged. Planned for 25 years hence, their alternative 'home' was designed predominantly by Alison Smithson, and made of a type of plastic capable of mass-production in a single unit. This 'gadget-filled dream of the middle-income household' included a self-cleaning bath, easy-to-clean corners and remote controls for the television and lighting.[84] Architects played an important part in the development of new design in Britain in the early 1960s, most obviously via Archigram, founded in 1961. Against 'gutless architecture', Archigram adopted an avant-garde stance, publishing a magazine and producing exhibition-cum-manifesto pieces, such as the Plug-in City, Walking City and Living City between 1963 and 1964, and the Instant City in 1968.[85] An exhibition in Harrod's in 1967 for a house for 1991 included moveable walls and floors with inflatable sleeping and seating structures.

Such avant-garde activity was not solely focused on London. During 1954 Richard Hamilton, a tutor in Fine Art at Newcastle University since 1953, had organized the *Man Machine Motion* exhibition, which opened at the University's Hatton Gallery in May 1955 before moving to the Institute of Contemporary Art (ICA) in London. A precursor to *This is Tomorrow*, the exhibition introduced Newcastle art audiences to a radical type of display,

'A mad mixture of pinks'
living-room in the *Design
Centre Comes to
Newcastle* exhibition,
Callers Furniture Store,
Northumberland Street,
Newcastle upon Tyne,
1969.

Radical exhibition design
by Helen Challen in the
*Design Centre Comes to
Newcastle* exhibition,
Callers Furniture Store,
1969. On the left, the chair
in white and gold was by a
student from Newcastle
College of Art and
Industrial Design.

although the extent to which this resonated with the local population was
debatable.[86] Photography rather than art, the exhibition revealed the obsession of Hamilton and collaborators with technology; as Banham put it:

> The basis of selection of the material exhibited was that each image
> should show a motion-machine, or similar piece of equipment, and a
> recognisable man. Photographic images were preferred since photog-
> raphy is more or less coeval with mechanised transport and belongs
> to the same technological environment.[87]

Designing Modern Britain

Teenage bedroom in the *Design Centre Comes to Newcastle* exhibition, Callers Furniture Store, 1969.

Wrighton & Sons kitchen in orange and white. Further right: dining table and chairs (Robin Day) in the *Design Centre Comes to Newcastle* exhibition, Callers Furniture Store, 1969.

Inspired partly by his weekly train journeys from Newcastle to London, Hamilton was fascinated by movement via technological means. In Newcastle, Hamilton was one of a number of contemporary artists who contributed to the post-war cultural landscape of the city and to the university's reputation as a centre for progressive art education. With Victor Pasmore, he had introduced 'Basic Design' as a way of teaching art. Pasmore, who at this point was Head of Painting, had worked on Peterlee New Town in the mid-1950s. His work was Constructivist in orientation, as represented in the Apollo Pavilion that he designed for Peterlee, with its

severe geometry, Brutalist concrete aesthetic and monochrome abstract quality.

Peterlee was one of 22 new towns that followed the New Towns Act of 1946, and it offers an insight into the inherent problems in using a standardized, modernist design language to provide new public-sector homes in a regional context, in this instance a close-knit working-class mining community in the Durham coalfield. The new town took its name from the self-educated miner's leader and Durham politician, but that was about the extent of its eventual engagement with local identities. In taking on the job in 1948, this had certainly not been Berthold Lubetkin's aim. He had gained a reputation in the 1930s with the design of the privately owned flats Highpoint 1 and 2 in Highgate, London, but fired by left-wing ideas, he designed Finsbury Health Centre (1935–8) 'to give the citizen of Finsbury a new deal'.[88] He believed that Peterlee offered an opportunity to design something quite distinct from Harlow and Stevenage, which were effectively overspills from London. This new town required something different, and after holding meetings with miners, Lubetkin aimed to produce a design equivalent to the solidarity, strength and shared culture of the mining communities, to provide, in effect, a 'mining capital'. The brief was to house 30,000 people, to provide a recreational and shopping centre, and to stop adverse demographic change – notably the drift away from the area by young women who could find neither jobs nor housing (much housing was tied to the mining industry). Lubetkin's plan, which took advantage of steep wooded denes, was

Designing Modern Britain

based on 'a densely developed town with a compact group of large and tall buildings in and around the centre', in contrast to the sprawl of new towns in the south-east of England.[89] Its plan resembled a Constructivist painting with abstract, geometric forms, and although Lubetkin had envisaged the design as 'helping the local community to realise its aspirations', after endless wrangles with the National Coal Board, civil servants and central government, he resigned in 1950, believing that policy decisions had eroded 'the original idea of a "bastion for miners" into any other new town'.[90] An attempt to improve the subsequent plan, described as 'a descent from the spectacular to the nondescript' by the Development Corporation's General Manager, came in 1954 with the aid of Victor Pasmore.[91]

Like Lubetkin, Pasmore was well versed in the use of Constructivist forms underpinned by modernist design theories, but a painter rather than an architect, his design was low-rise and flat-roofed, 'a transposition of Mediterranean light and line, a dialogue between art, architecture and urban design'.[92] In a radio interview in 1967, Pasmore claimed:

> Housing and road layout have been treated as different organic processes and they have been oriented in such a way as to allow both factors to complement each other through opposition on one solid and rigid; the other linear and elastic.[93]

In the same year, Pasmore designed the Apollo Pavilion to act as a focal point in the new town. Under threat of demolition in the late 1990s, this has since been described as 'the only surviving element embodying the idealism that once informed Peterlee New Town and, in national and international terms, is a rare example of a truly spatial creation, crossing the boundaries

House of the Future designed by Alison and Peter Smithson, exhibited at the *Daily Mail Ideal Home* exhibition, 1956.

of art and architecture'.[94] Both Lubetkin and Pasmore's interventions failed to achieve a synthesis of post-war modernist planning and design with distinctive regional identities. In fact, community demands from the 1970s necessitated the flat-roofed standardized housing being given a vernacular face-lift (pitched roofs) along with upvc windows to guard against the north-east winds and rain.

A development that paralleled Peterlee took place in Newcastle upon Tyne in the 1960s. Planning, a feature of wartime, was carried on into the post-war period in many of Britain's cities, particularly those that had been bomb-damaged, such as Coventry, Sheffield and London. Newcastle had not been devastated by bombing, but had been subject to several grand planning schemes. A flurry of plans culminating in the Development Plan of 1963 set the city on a path to reconstruction, not only in terms of retailing, business and cultural activities in the city centre, but also in relation to housing in the inner suburbs to the west and east. The vision for Newcastle by T. Dan Smith, leader of the Council from 1958 to 1966 (and subsequently notorious for political corruption), was embodied in the 1963 plan.[95] The plan had three main

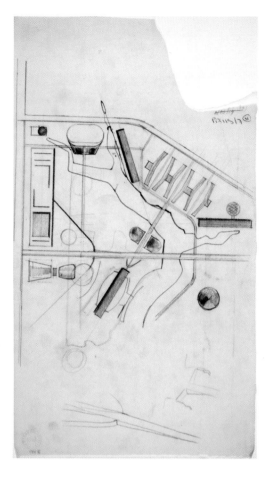

Constructivist Plan for Peterlee New Town designed by Berthold Lubetkin, c. 1948.

elements: first to develop the road system to deal with the increased amount of traffic and the bottleneck over the River Tyne; second to renew the city centre; and third to develop approximately 25,000 new houses and to revitalize 12,000 others. Visually striking were the plans for the city itself: combining underpasses and walkways, urban motorways, tall blocks interspersed with medium-height asymmetric office and residential developments and retailing centres, it was the embodiment of Le Corbusier's pre-war vision of a modern city.[96] Much of it was built with the aid of well-known architects such as Basil Spence, who designed the new Central Library and All Saints' office development; Robert Matthew, who was responsible for the Swan House development, and Arne Jacobsen for the Eldon Square Hotel (not built).[97] In addition, areas of former working-class housing were demolished to the east and west for council housing, and new areas developed to the north; thus 'the planning strategy for the city's

Designing Modern Britain

residential areas was built around the concept of neighbourhood units', including Shieldfield and Byker to the east, and Scotswood to the west.[98] These developments were based on high-rise-system building techniques, delivering not only large quantities of housing, but just as important, consumers for the new fridges, cars, vacuum cleaners, washing machines and televisions on which post-war Britain's prosperity depended. Newcastle, however, remained economically 'fragile', mainly because of the processes of deindustrialization that gathered pace after the war. Disguising this was its dramatic growth as a regional retailing centre, which saw retailing turnover in the central shopping area of Newcastle increase from £56 to £121 million between 1951 and 1971.[99] Large retailers such as Marks and Spencer, Littlewoods, British Home Stores, Woolworths, Burton, Callers Ltd, and the department stores Fenwick, Bainbridge and Binns, catered for the new categories of desired goods. At the same time, Newcastle, like other cities in Britain, began to develop a distinct retailing infrastructure based on youth cultures, particularly small independent boutiques.

Newcastle upon Tyne's urban motorways were indicative of the transformations brought by the expansion and proliferation of car cultures on city planning and everyday life in post-war Britain. The motor car had been a potent symbol of modernity for early modernists such as Le Corbusier, who, in the 1920s, celebrated its technological innovation, machine aesthetic and Fordist production in his promotion of a modernist aesthetic in architecture and design. Later theorists such as Herbert Read in *Art and Industry* (1934) and Roland Barthes in *Mythologies* (1972) recognized the significance of the motor car in contemporary visual culture, the former comparing the technological advances of bridge construction with that of a motor car, the latter likening the new Citroën DS (shown at the Paris Motor Show in 1955) to a cathedral. In Britain, there was dissonance between the glamour of the motor car and the reality of its production. As Paddy Maguire has argued, Fordism remained unattainable and unevenly implemented across British industry in the 1950s, and this was also the case with the car industry, since craft methods persisted and model competition increased.[100] But technological innovation played an important part in the development of cheaper, smaller cars in the post-war years, especially the Morris Minor, the Morris Mini and the Morris 1100, which went on sale in 1948, 1959 and 1962 respectively. All three cars were designed by Alec Issigonis for Morris

Housing at Peterlee, c. 1951.

The central motorway, Newcastle upon Tyne, after the 1963 plan.

Motors, founded in Oxford in 1912. An engineer by training, he brought an essentially modernist approach to car design, rethinking basic elements so as to produce economical, easy-to-drive and visually distinctive cars for a mass market. Indeed, the Morris 1100 also benefited from external styling by Sergio Pininfarina's Italian design studio. The organic streamlined aesthetic that emanated from the Pininfarina studio had a significant influence on car design in Britain, especially for the top end of the market, exemplified by the E-type Jaguar available in the early 1960s. But in 1950s Britain it was the US automobile, in particular Detroit styling, that attracted most critical attention. For the Independent Group, planned obsolescence – summed up as 'short-term, low-rent chromium utopia' by Richard Hamilton – was the antithesis of the COID conception of good design, and thus had distinct appeal to those entranced by mass and popular cultures.[101] The extremes of US car styling were not adopted in Britain, although Vauxhall, the British subsidy of the General Motors, introduced transatlantic styling – evident in panoramic windscreens and white-walled tyres – with the Victor and Cresta in 1957. Influences from Italian design were discernible at both the exclusive and the cheaper end of the personal transport market, with the introduction into Britain of Piaggio's Vespa and Innocenti's Lambretta scooters in the early 1950s. These were technologically innovative, stylish (the mechanical elements were encased) and easy to manoeuvre. Targeted at the young and women, the scooter was disparaged by the stalwarts of the British motorcycle industry as effeminate and unmanly. But to these emerging markets, Italy epitomized chic, modern design for a youthful urban lifestyle that was encapsulated in the design of the scooter.

Designing Modern Britain

Resistance and Incorporation: Being Fashionable in Post-War Urban Britain

Threading through Dick Hebdige's account of post-war subcultural style in Britain is the contingent relationship between the first and subsequent generations of West Indian immigrants, arriving in Britain from the late 1940s into the 1960s, and white working-class communities.[102] Wearing lightweight, pale, tweed or mohair suits with dazzling picture ties and pork-pie hats,

Morris Minor, designed by Alec Issigonis, 1948.

Vauxhall Cresta, 1957, showing features drawn from the US car industry such as white-wall tyres and wrap-around windscreen.

it was an ensemble so sharp that these purveyors of style appeared to slice their way through the smog of Britain's major cities. It was a potent, capricious mode of dress worn en masse by black, working-class immigrants.[103]

Occupying the same geographical spaces as their white working-class neighbours, early groups of West Indian immigrants were confirmed Anglophiles . . . they shared the same goals, sought the same diversions . . . and despite the unfamiliar accent, drew upon the same 'language of fatalism', resigned to their lowly position, confident that their children would enjoy better prospects, better lives.[104]

It was with their children, for whom the better life did not materialize, that challenging and questioning gathered pace and became more assertive; 'second generation black British youths were having none of this, no longer wanting to be called British, considering themselves West Indians, considering themselves black.'[105] Music, style and language offered a means to question dominant definitions of black identities. The various sub-cultural stances developed by black youths in the 1950s and '60s influenced white working-class groups, although there were occasional convergences between black and white, particularly through music – two-tone and ska were examples. As American popular culture gained popularity, 'teenage life began its gradual move to centre-stage'.[106] In 1960 British teenagers spent £850 million largely on themselves, buying records and record-players, cosmetics, clothes and cinema tickets; this represented 5 per cent of the gross national spend for that year.[107] An exemplar of white teenage consumption was the Teddy Boy, who 'visibly bracketed off the drab routines of school, the job and home by affecting an exaggerated style which juxtaposed two blatantly plundered forms (black rhythm and blues and the aristocratic Edwardian style)'.[108] If black identities – via immigrant West Indian communities and black urban America – helped to shape emerging youth identities in post-war Britain, then so too did class. Resolutely working class, the Teddy Boy drew elements of style from his social superiors, the New Edwardians, and in this reaffirmed, perhaps even parodied, a class-specific monied 'English' identity. These new youthful sub-cultures were located temporally and geographically in the city and suburbs of London and other large cities, but they were attuned to wider influences from Italy, the USA and the West Indies. Their sartorial acquisitions sometimes required in excess of £50, as good tailoring did not come cheap. The 'style', comprising sharply cut suits with slim drape jackets, bootlace ties, suede shoes ('brothel creepers') and elaborate quiffs and sideboards, all heavily Brylcreemed, lasted until the late 1950s. The Teddy Boy has assumed an iconic significance in the story of youth cultures in post-war Britain: 'Teddy Boys were the start of everything: rock 'n' roll and coffee bars, clothes and bikes and language, jukeboxes and coffee with froth on it – the whole concept of a private teen lifestyle, separate from the adult world.'[109] In the light of Hebdige's insistence on the centrality of black cultural identities (both West Indian and American) on youth cultures in Britain, this overstates their importance. But they were an example of

Young man wearing Zoot suit, 1960.

identities in 'the process of transformation': becoming modern, not just in style, but in terms of attitudes and lifestyle; peering beyond the terraced streets and glimpsing foreign cultures, but also parodying entrenched British upper-class taste. These developed in response to economic, social and cultural change – full employment, lack of familial responsibilities, and burgeoning retailing systems – in this they contributed to the modernization and democratization of fashion and style that dominated fashion from the mid-1950s.

A characteristic of the relationship between youth cultures and fashion in the 1960s was its commodification, and in this respect Mary Quant was especially significant. A graduate of Goldsmiths College of Art and one of a number of designers in the 1950s and '60s who were art-school trained, Quant had established her first boutique, Bazaar, in 1955 on King's Road, Chelsea – an area that was still shabby, but with bohemian connections. Part of a largely middle-class set, she drew on the business acumen of her husband at the foundation of her business. Trained in illustration, she was famously ignorant of the processes of manufacture at the outset, and her early designs, violating traditional notions of good taste by mixing large spots and checks, attracted hostility from within the fashion trade, as well as being criticized for being poorly made. Described as 'simple, neat and unfussy, often using cotton gaberdine, poplin or gingham', the 1950s Quant 'look' quickly became popular.[110] She thrived partly because her designs sold to a booming youth market that took to the novelty of 'boutique window as pop-art installation'.[111]

By 1961, with the establishment of her second Bazaar in Knightsbridge, Quant had by necessity put her rather haphazard methods into some sort of order, initially wholesaling her designs and then in 1963 launching the Ginger Group to franchise these for mass-production. Although the predominant image of Quant's early activities was of someone who was rather dilettante, she had effective financial backing and an unerring instinct for breaking rules, whether in terms of clothing design, boutique display or fashion promotion. She also benefited from the structural changes in fashion retailing and marketing that had been established in the 1950s. The clothes were not cheap (a dress might be between £6 and £9), although this can be hard to judge, especially since the fashion-conscious young might do without in order to afford them, and since many still lived at home with their parents, even working-class girls could save up.[112]

Quant claimed that the Mods were an important spur to fashion in the early 1960s, but by 1963 the meaning of 'Mod' had shifted. Initially largely descriptive of a small male minority cult, who, understated and subtle in comparison to Teddy Boys, wore sleek Continental – particularly Italian – tailored suits, it came to mean instead white male and female 'pop' fashions.[113] Inevitably, the desire to be a truly fashionable teenager in Britain in the 1960s required a visit to Quant's shops in Chelsea and Kensington or to John Stephen's boutique in Carnaby Street, but as youthful styles spread beyond the boundaries of a handful of London streets and suburbs, more of the young knew of and were able to engage with Mod cultures. Illustrated in magazines (*Queen* and *Nova*) and newspapers, and visible via television (epitomized by John Bates's designs for Diana Rigg in the TV series *The Avengers*), film and pop music, these designs represented the height of modernity. The 'look' was minimal, with simple, hard lines, dropped waistlines as seen in pinafore and shift dresses, cut-away armholes, bright bold colours and simple dramatic patterns. Ostensibly an icon of sexual liberation, there was an ambivalence about these youthful images in which girls remained girls, not women.[114] The waif-like adolescent appearance of these 'girls', while defying one patriarchal construction of femininity, merely articulated another, which was innovative visually, but still dependent on male admiration.[115] Importantly, these contingent representations, shaped by shifting categories of race, gender and class, and initially London-based, stimulated youth cultures beyond the capital. By the early 1970s northern cities such as Newcastle and Leeds had their own boutique culture tucked away around the corner from the main shopping thoroughfares – Bus Stop on Northumberland Place in Newcastle upon Tyne and Boodle-Am in the Queens Arcade in Leeds. But these early 1970s boutiques heralded different forms of design culture, at once less singular and authoritative, rather more fragmentary and diffuse, retrospective rather than forward-looking and emerging, as Fredric Jameson put it, 'as specific reactions against the established forms of high modernism'.[116]

Ultimately, an official 'high' modernism was consolidated if not entrenched in crucial areas of design in Britain from the early 1950s to the late 1960s. Organizations such as the COID were central to this, although the rhetoric of modernity – progress, planning and opportunity – permeated the language of those politicians, critics, designers and artists who subscribed to a vision of a rational, forward-looking Britain. The COID attempted to promote the idea of design as a set of common objectives, although it became increasingly diverse towards the end of the 1960s as different consumers redefined and reinterpreted being 'modern'. In such a context, regional identities were equated with retrospection, and the processes of modernization as witnessed in Newcastle involved a full-scale adoption of 1920s European modernism by way of the COID. Equally, Lubetkin's experience at Peterlee spoke volumes in this respect, since he recognized that regional distinctiveness borne of shared community experiences counted for very little against the juggernaut of central government organization – including the blithe disregard of the National Coal Board, an early participant in the processes of de-industrialization that ravaged the north-east of England in the 1970s and '80s. When Festival of Britain planners addressed the north of England, Scotland, Northern Ireland and Wales as separate and distinct, regional differences were collapsed into stereotypical references to history, tradition and the countryside ('highland' games in Scotland and hill farming in Wales). The promise of cohesion flagged up in Festival brochures was probably unattainable, undermined by the international, global nature of the processes of modernity, economics and consumer markets. Subject to new ideas from abroad (the United States and Europe in particular) and at home (immigration from the West Indies especially), design in Britain was nevertheless in the thrall of modernist rhetoric for a substantial part of the period under discussion. But by the late 1960s, the failures of modernism in solving crucial design problems were being glimpsed, and any sense of coherence was glaringly exposed by an increasingly fragmented society. Immigration from the Indian subcontinent, youth protest, women's liberation and gay rights emphasized the growing diversity of cultural life in Britain, and its attendant design.

THE BEATLES IN PEPPERLAND

Above: the *Eleanor Rigby* sequence

Above: Pepperland. Below: a Meanie

As John and Paul sing *All You Need Is Love*, the words leap from their lips and gradually fill the screen . . . one of several highly imaginative episodes in *Yellow Submarine*, which opens at the London Pavilion on Wednesday. Described (a shade pretentiously) as "a mixed-media feature – live, animated and something quite else", the film is in fact a full-length (about 90 minutes) cartoon in which the Beatles appear as the central characters. A fantastic and discursive story culminates in the heroes liberating the inhabitants of Pepperland (by the simple means of song) from the tyrannical oppression of a race of Meanies. The plot is essentially a device to link a dozen Beatle numbers, of which eight are old favourites and four are new : *Northern Song, Hey, Bulldog!, It's All Too Much* and *Altogether Now*. The whole thing has been designed by Heinz Edelmann, the German illustrator. Like such characters as Mickey Mouse and Donald Duck, the Beatles appeal in different ways to audiences of every age. More sophisticated than *Snow White*, but equally inventive, *Yellow Submarine* just might repeat the Disney success of over 30 years ago.

5

The Ambiguities of Progress: Design from the Late 1960s to 1980

Retrospection and eclecticism permeated design from the late 1960s through the 1970s, and this was evident in the growing enthusiasm for visual styles from the past (Arts and Crafts, Art Nouveau, Art Deco), second-hand clothes and antiques (particularly from the late nineteenth century and before the Second World War) and non-Western cultures (India especially).[1] In 1968 the *Sunday Times Magazine* featured a brief account of the Beatles' new cartoon film, *Yellow Submarine*, in which the 'Fab Four' donned fluorescent 'retro' military-style jackets; it reviewed the Paris fashion collections in August that had included a Cossack-style pantsuit from Patou alongside a Courrèges short cape and visored hood; it ran a piece on Ginger Rogers showing stills from various 1920s and '30s films; and its 'Design For Living' section included an article showing models posed on oriental patterned rugs wearing 'Eastern-inspired' fashions.[2] Indicative of increasing disillusionment with 'official' or high modernism, by the early 1970s a process of critical questioning was underway, as it became apparent that the brave new world promised in the 1920s and '30s and comprehensively planned after 1945 had been implemented only in part and with limited success. In fact, during this period a number of counter-cultural stances emerged that questioned central modernist tenets, including the uncritical commitment to progress, the enthusiasm for technology and the belief in social advancement and equality through design. Some believed that far from being radical, modernism had in fact become little more than a style; and that mass-production had delivered the 'opiate' of mass consumerism, rather than a universal design culture. The 'modern' had been effectively deployed to persuade the consumer to buy into particular 'lifestyles' and different 'identities'. In this way, rather than eroding social inequalities – a central aim of European modernist theorists in the

Feature in the *Sunday Times Magazine*, 14 July 1968.

161

1920s – design was reproducing and embodying them. At the same time, there was a growing recognition of disparate political voices that began to prove surprisingly effective in articulating a wider sense of discontent. Civil rights protests in the USA and the assassination of Martin Luther King in Memphis in April 1968, student riots in Paris in the same year, the emerging women's movement, anti-Vietnam marches and demonstrations, such as at the US embassy in Grosvenor Square in London in March 1968, and green politics were widely reported via the media. This paralleled and contributed to a period of sustained debate in the wider culture about the role of design. Influential was Victor Papanek's *Design for the Real World: Human Ecology and Social Change*, published in 1971, which proposed a different understanding of what design entailed.[3] There was growing recognition that design was not merely about satisfying (and stimulating) consumer desires, but that it had wider social, ecological and ethical consequences. This was reinforced by the oil crisis of 1973–4, which highlighted the finite state of world resources and exposed the dependency of highly developed Western economies on hitherto cheap energy and raw materials. Green politics began to figure more prominently from the 1970s, demonstrated by the emergence of retailers such as The Body Shop in 1976. The question of ethical trading also began to be debated, and organizations such as the Christian-inspired company Traidcraft – begun in 1979 – aimed to 'raise awareness of issues relating to poverty among consumers in the UK, and encourage them to make moral choices as they spend their money'.[4] Such a stance was in marked contrast to the promotion of global brands and the norms of world trading that prioritized political and economic factors, rather than ethical ones.

Consumerism was a critical driver of the post-war economy in Britain, and design played a crucial part in this by encouraging people to buy new, modern goods. More often, things were bought not as necessities but as representations of lifestyle aspirations, and manufacturers, retailers and designers became adept at creating and manipulating consumer desire. Abundance, aspiration and desire underpinned the design system in the West, and inherent in this was the idea that design was a social language, and that it expressed lifestyle.[5] The consumption of goods, as opposed to the production of goods, began to define the cultures of design from the late 1960s onwards. Shops and shopping were central to the articulation of design as the crucial element of 'lifestyle': this was typified by Biba and Habitat, and although designers such as Barbara Hulanicki and Terence Conran played important roles, it also became apparent that they were part of larger complex design systems that involved advertising, marketing and promotion. At the same time, consumers became more knowing and discerning in their choices as advertising, magazines and eventually television

Designing Modern Britain

proliferated. As the dominance of modernism was undermined, it was no longer possible, or perhaps even desirable, for one particular set of values or a distinct approach to claim unique legitimacy.

Fashion, seemingly the ultimate consumerist vehicle, exemplified the multiplicity of styles, attitudes and approaches. Counter-cultural and oppositional, but at the same time intrinsic to the late capitalist economy, a number of 'looks' coexisted: hippies, glam rock, punk, 1920s and '30s retro, as the past was systematically plundered for new ideas. Dubbed 'Style Wars', it was a product of 'market segmentation', but at the same time it epitomized a shift from modernity to postmodernity.[6] If the modernist vision was positivistic, technocratic, rationalistic and based on a belief in linear progress, absolute truths and standardization, then postmodernity was characterized by heterogeneity, difference, fragmentation and 'an intense distrust of all universal or "totalising" discourses'.[7] This might be thought to sum up official or high modernism, but as this book has shown, modernist theories and practices were plural, not singular, especially when synthesized with 'English' or 'British' design traditions and changing subjectivities. Design in the 1970s and early 1980s contributed to and reproduced 'widespread individualism and entrepreneurialism in which the marks of social distinction were broadly conferred by possessions and appearances', but it was also subversive.[8] Prescient of 1960s optimism, rationality and modernity, Macmillan's phrase, 'you've never had it so good', could be seen by the end of the 1960s and early 1970s to be not only over-optimistic, but naive.

What Went Wrong? Being 'British' in the 1970s

Elizabeth Roberts's oral history of post-war Britain identified a recurrent question put by several respondents: 'why had things gone wrong'?[9] With no obvious answer, she pointed out that whereas there was greater certainty and predictability about aspects of everyday working-class life before 1940, 'in the post-war world ever-expanding choices rendered such certainties and generalities dangerous.'[10] Her study – examining attitudes and behaviour in working-class life through the voices of individual witnesses in the north-west of Britain – provided real insights into social and cultural change over time. Focusing on women, it nevertheless highlighted broader issues affecting their lives, and it assessed the wider and the individual impact, for example, of the dramatic increase in married women's employment – one of three factors identified as being indicative of major change since 1945.[11] A central concern of her study, however, was the varied perception of these changes, whether they were considered good or bad, and the extent to which this had contributed to a sense that something had gone wrong. Pointing out that by 1970 many had achieved 'a house they owned, a comfortable income, and a secondary and possibly higher education for their children' – all symbols of middle-class status and success – Roberts doubted that anyone could suggest that these were 'bad times'.[12] She concluded, however, that such questioning pointed to 'the ambiguity of progress', and reiterated the views of other historians about the difficulties of assessing such qualitative perceptions. In an attempt to judge perceptions more quantitatively, some have identified a growing awareness by the end of the 1960s that Britain's economic growth was not delivering as much as had been expected, nor was it performing well against other developed nations, particularly the former enemies Japan and Germany. Several historians argued that the rise of inflation and unemployment, the deterioration in industrial relations and the financial problems experienced by the public and private sectors (for example, in the Welfare State, and in shipbuilding and aerospace) contributed to the view that the period of consensus and stability begun in the mid-1950s had come to an end.[13] Others proposed that the desire for consensus stemmed from a determination to escape for ever from the Depression conditions of the 1930s, but that 'the story of the middle to late '70s might well be seen as one of a return to the gloom of that "devil's decade"'.[14] The oil crisis of 1973–4 (which followed the Arab–Israeli war and led to oil shortages and a fourfold price increase), strikes by local authority manual workers, dockworkers, railway workers and miners, and the 'three-day week' marked out the early 1970s as an especially troubled period. The stark economic facts were that wages fell in the mid-1970s, after progressively increasing from 1950, and Britain's traditional manufacturing industries were in continuing decline. The latter in particular contributed to regional inequalities, and by the end of the 1970s and early 1980s, as in the 1930s, structural shifts in the British economy could be mapped on a 'North–South' divide, which some characterized as lying between 'a prospering innovative South (the Midlands and everything to the south, but excluding Wales), and a backward, depressed North (everything north beyond the Midlands)'.[15] Ensuing high unemployment rates and a mismatch between jobs and skills (more jobs in the service sector, fewer in the industrial sector) affected specific groups of workers – particularly young working-class men – and 'by the beginning of the '80s a job was becoming again, as it had in the 1930s, something that you began to thank your lucky stars you had'.[16]

There were anxieties about the social changes that had taken place too, particularly about the impact of permissive legislation and attitudes in society at large. A dramatic increase in the divorce rate, particularly in the

Designing Modern Britain

1970s and '80s, and the rise of illegitimacy beginning in the 1960s was perceived to contribute to the breakdown of the 'traditional' family and the steady increase in one-parent families. But the 'ideal family' was only ever that – an 'ideal' – and family structures had been in transition throughout the twentieth century, caused by war, death, work and changing gender relations. New legislation such as the Abortion Act of 1967 and the Divorce Act of 1969 had contributed to this permissiveness, but feminists recognized that sexual equality was still a long way off, and there were concerted campaigns for women's liberation and social equality. During the early 1970s a plethora of publications contributed to a fundamental analysis of the position of women in Britain, by women historians such as Sheila Rowbotham and social scientists such as Elizabeth Wilson, but at a popular level it was Germaine Greer's *The Female Eunuch* of 1970 that raised wider awareness. Other important legislative change took place, including the Equal Pay Act, which, passed in 1970, did not come into force until 1975, and the Sex Discrimination Act of 1975. Legislation was slow to have impact, and there was growing awareness that greater equality was dependent on cultural change as well as political and economic factors. In this respect, the cultural arena came under scrutiny and certain aspects of design, such as domestic product design, housing, advertising and fashion, were emphasized. Although many of these – fashion by way of an example – could appear intensely patriarchal and confirming of the status quo, there was growing recognition that design was a visual language that contributed to the formation of identities – gendered and otherwise. Such thinking and action were part of a broader process of political questioning that gathered pace from the late 1960s and '70s, focusing on young people, women, black and Asian groups, who were disenfranchised in a variety of ways. This 'fractured consensus', combined with an acknowledgement that progress was not straightforwardly 'good', began to penetrate people's everyday experience from the end of the 1960s through the 1970s – hence the questions from Elizabeth Roberts's oral-history respondents. Some historians argued that optimism and a commitment to progress – 'the optimistic consensus' – had 'successfully carried Britain through the difficult post-war years into the affluence of the '60s', whereas others suggested that this had 'concealed the desperate realities of Britain's true predicament'.[17] Whichever view was correct, it was increasingly the case that 1970s governments, whether Conservative or Labour, did not have the solution, and that 'all countries suffered a substantial slowdown of productivity growth following the oil crisis, though Britain continued to perform significantly worse that her major European rivals and Japan'.[18]

By the end of the 1960s, and certainly during the 1970s, it became more obvious that economic affluence and prosperity were not shared equally and that some sections of British society were economically deprived, politically disenfranchised and socially marginalized. Indicative of this were the experiences of black communities, but, as during the Miners' Strike in 1984–5, these were not totally disempowered because communities developed alternative, informal lines of organization and resistance, often employing diverse strategies that used popular cultural forms often rooted in design, such as fashion and graphics. Nevertheless, during the 1960s and early 1970s both Labour and Conservative governments had restricted immigration by a number of legislative means, while at the same time enacting anti-racist legislation. These restrictions in effect condoned racism: 'The practice of racism by the state in keeping out black people has only served to legitimise racism in society as a whole.'[19] As historians and economists have pointed out, 'black people in Britain suffer from more unemployment, poorer housing, education and health care and have little mobility within the employment market.'[20] In addition, following the Immigration Act of 1971 and the increased activities of the National Front (putting up 54 candidates in the General Election of February 1974), and perhaps taking a lead from Martin Luther King, who visited London in 1965, resistance to racism began to be organized by black communities, often focused on local, informal networks.[21] Significantly, this often took place outside formal politics:

> It relied for its development on networks of culture and communication in which the voice of the left was scarcely discernible and it drew its momentum from the informal and organic relationship between black and white youth which sprung up in the shadow of 1970s youth culture.[22]

The 'Rock Against Racism' action, for example, showed the ways in which culture could be both political and genuinely transformative. Youth cultures and music were important sites, and in their writing Gilroy and Hebdige made explicit 'the hitherto coded and unacknowledged relationships between black and white styles'.[23] As well as providing opportunities for resistance to modernist hegemony and consumerist aspirations, cultural exchanges of this type, sometimes between the disparate and the marginal (black and white youths in this instance), were indicative of postmodernism. Indeed, Roberts's 'ambiguity of progress' and Marwick's 'fractured consensus' were also symptomatic of this.

Modernist Dystopias

The architecture critic Charles Jencks famously dated the end of modern architecture to 15 July 1972, when the Pruitt-Igoe housing scheme in the American city of St Louis was blown up. Built between 1952 and 1955, it had become a symbol of the failure of the modernist vision of design as social engineering, a tool for creating a better, fairer, rational society. As Jencks put it: 'Good form was to lead to good content, or at least good conduct; the intelligent planning of abstract space was to promote healthy behaviour.'[24] By the time of its demolition, the scheme had been vandalized, defaced and rejected by its largely poor inhabitants, although it had won an award in 1951 for its design, which incorporated many modernist elements: 'streets in the sky', new technologies in construction and design, rational planning and standardized 'egalitarian' housing units. Using this as an example of the failure of modernism, Jencks described a new way of thinking about architecture as 'radical eclecticism', or 'multiple-coding', which in Britain involved the intentional deployment of incongruity and transience, the use of ad hoc and pop imagery as well as visually striking structural innovations.[25] The building of high-rise mass-housing schemes continued throughout the 1970s, however, and Britain had its own versions of Pruitt-Igoe as the public lost faith in mass housing and blocks were demolished from the late 1970s.[26] Modernism's most public failure, however, was in 1968 in east London, when a 23-storey tower block, Ronan Point, collapsed due to inadequate construction methods, killing four inhabitants. By this date numerous modernist tower blocks and mass-housing schemes had been built, with many still planned. In 1961 the Labour leader Hugh Gaitskell opened the local-authority-built Park Hill flats in Sheffield, one of Britain's largest modernist housing schemes. Influenced by Le Corbusier's Unité d'Habitation in Marseilles, it was one of the most spectacular examples of new approaches to communal living in post-war Britain. Huge in scale and ambition, it occupied a hillside overlooking Sheffield city centre, providing almost 1,000 flats housing more than 2,000 people. With deck access, it attempted to preserve the communal benefits of street life. Ironically, but perhaps unsurprisingly, due to the scale of ambition, it was in the context of such housing that the failures of modernism were most apparent. Although many architects were inspired by the tenets of modernists such as Le Corbusier and keen to work in this way, they often had little choice but to use new (often untried) technologies in order to build quickly to provide for new homes in the aftermath of war, but also in response to political pressures, since housing the working class became a crucial political battleground in the post-war years. Park Hill, like Killingworth New Town in Northumberland built a year later, aimed to provide a better way of life, but instead became a site of social disintegration, since old established working-class communities failed to adjust to living in accommodation that lacked, among other things, private entrances, enclosed gardens and some measure of individuality.

A special issue of the *Sunday Times Magazine* on 28 March 1965 focusing on 'The North' looked in particular at the north-east of England, with some additional references to Yorkshire and the north-west.[27] In this, Newcastle upon Tyne and surrounding towns – Durham, Sunderland, Teesside and Hartlepool – were used as metaphors for post-war renewal, and in a number of interviews with local inhabitants and incomers from the south-east the focus was on the wider cultural and social amenities of the area. Emphasis was put on traditional ways of life and experience, but this was tempered by a recognition that the region was changing. Housing, retailing and pop cultures were cited as evidence of this, and amidst a sequence of black-and-white photographs depicting stereotyped images of northern life – nineteenth-century terraced streets, women hanging out the weekly wash, dirty miners and pit ponies, young women with curlers eating chips from a bag – a handful of new images highlighted the region's modernity. These included La Dolce Vita Club on Lower Friar Street in Newcastle, where Ella Fitzgerald, Sarah Vaughan and Manfred Mann performed; modernist abstract housing designs by Victor Pasmore at Peterlee New Town; the Northern Gas Board headquarters at Killingworth New Town; and new local authority housing redevelopments on the Scotswood Road in Newcastle. Written at the height of post-war modernity, its symbols were poignantly those of design and architecture. Killingworth was praised for its planning, particularly the separation of traffic and pedestrians, its picturesque moat-cum-reservoir, and its striking new office development with fibreglass tower. Peterlee, providing homes for workers moved north from the south-east of England with office and factory relocations, was described as 'decoratively austere and white-painted shapes', but it lacked essential new technologies: central heating and double glazing.[28] The new Newcastle city centre was cited as evidence of an outward-looking approach, especially the employment of architects with national and international reputations. But it was to the leader of the council, T. Dan Smith, that the greatest achievements were attributed, the 'redevelopment of the notorious Scotswood Road slums'.[29] With the exception of the community centre, which was designed and landscaped by members of Newcastle University School of Architecture, the ten tower blocks of between 12 and 15 storeys soaring above the River Tyne on the Scotswood Road were designed by the

Newcastle City Architects' Department, and were instantly evocative of Le Corbusier's City of Towers illustrated in *Vers une architecture* in 1923.

At odds perhaps with the apparent modernity were the local people, who were characterized as warm, friendly and trusting, but essentially provincial. An incomer from the south-east remarked that he had yet to see a slag-heap, but 'the main disadvantage is an urban one – style'. Describing La Dolce Vita as ' a bit cheap and brassy really, like a big sort of 1940s dance floor with coloured lights underneath', he commented: 'In London, whatever you've got, you've got the best of it. But here you've got the other side of the coin, you participate.'[30] Equally, the 19-year-old daughter of another family relocated to Peterlee noted:

> It's 20 years behind the fashion . . . There's nowhere nice to dance. And some of the boys – I was walking down the street and I nearly dropped dead. Those tight jeans, and long jackets – *Edwardian* things. And long hair, not like now, but like it used to be. *Side-boards*. Its all rockers, no mods.[31]

Modern urban life, of the type that might be experienced in London, could be found in the region to a certain degree, but in Newcastle upon Tyne, the principal city, certainly not in Peterlee.

While the 1960s did witness increased prosperity in the city, regional enthusiasm for modernity evident in the special issue of the *Sunday Times Magazine* had been dented by the late 1960s and '70s as corruption engulfed the principal exponent of regional regeneration, T. Dan Smith, and as elements of the early 1960s urban plans began to be questioned. Urban motorways, which cut communities in half; New Town developments with dismally inadequate housing; tower blocks with little sense of social cohesion; the wholesale demolition of the decent terraced housing in older working-class areas of the city, as well as the loss of the late Georgian and early Victorian commercial and residential terraces and squares (from the Grainger–Dobson era) were indicative of the inadequacy of modernism. As late as 1969, important historic buildings were still being demolished in Newcastle, despite local opposition and the activities of the Georgian Group and the Royal Fine Art Commission, although retrospectively it is apparent that the loss of early nineteenth-century buildings such as Eldon Square and the Royal Arcade did contribute to a rethink about the advisability of further new development.[32]

The growing awareness of the value of historic buildings was paralleled in Newcastle with the development of new approaches to the design of public housing. The city provides a very useful case study of the rejection of modernism in favour of greater diversity and a more locally responsive approach to design, exemplified by the building of the Byker Wall, a local authority housing scheme designed by Ralph Erskine. Begun in 1969, when Erskine took the extraordinary step of setting up office in one of Byker's old buildings, and largely completed by the mid-1970s, Byker marked a change in both the design of public housing and the ways in which architects communicated with clients. Described as an example of postmodern 'bricolage', Byker is a mixture of low-rise blocks of housing contained by a large 'wall' of maisonettes.[33] The visual style of the housing was vernacular, referencing pitched roofs, natural wood and brick. It was constructed in concrete, but various shades of red-and-brown facing brick added richness, and brightly coloured timber detailing contributed to a feeling of individuality. The overall plan was organic rather than rigidly rectilinear, with extensive planting throughout the site. Unusually, the architect aimed to produce a design that evoked 'local' individuality – to maintain 'valued' traditions and the specific characteristics of the neighbourhood – and he planned to re-house those already resident in Byker without breaking family ties and other established connections and ways of life.[34] The approach that Erskine took was antithetical to modernism in that it aimed not to create a new type of 'rational' inhabitant, but instead to respect existing communities and ties. It drew on 'local' vernacular styles, and although using modern methods of construction, incorporated traditional materials. One of the largest housing developments in Britain, it was nevertheless relatively human in scale with a mixture of housing types (as Jencks would have it, a combination of 'radical eclecticism and multiple-coding'),[35] and became one of the best examples of a postmodern approach to housing in Britain.

Consumption and Cultural Capital: Promoting Lifestyle

Within the framework of the COID, modernist approaches to design were mainly interpreted in relation to the rational use of materials, a thorough understanding of function and a concern for formal simplicity. But there was a shift away from this by the early 1960s, and 'being up-to-date and fashionable was often more important than being functional'.[36] Design became an extension of the fashion scene, perceived as a crucial element of lifestyle and as a means of self-expression:

> More than ever before, the effects of the press, the radio, television and the cinema entered the lives of practically every individual in the industrialized world, providing new sources of information, creating new expectations and suggesting new values. Inevitably, within this changing cultural climate, design diversified, taking on new guises and performing new roles.[37]

An array of newspaper features, magazines, practical manuals and books advised the consumer about design; as well as being informative (what to choose, where to buy from, etc), this literature also emphasized the broader significance of design, aiming to educate the consumer in the nuances of its meanings.

At a popular level were the colour supplements of the Sunday newspapers, particularly the *Sunday Times*, the *Observer* and the *Sunday Telegraph*. Visually striking, these introduced pop graphics, high-quality colour photography and an investigative journalistic approach to a wider readership. The *Sunday Times*, for example, was taken over by Lord Thomson in 1959, and subsequently it developed both a strong innovative visual style and a reputation for social campaigns based on investigative journalism, such as its exposure of the notorious slum landlord Peter Rachman; its role in the campaign for compensation for the British victims of the Thalidomide drug; and, on the cusp of second wave feminism in September 1971, a special issue headed 'All about Eve'. At its introduction on 4 February 1962, the magazine's typeface was adapted from Herman Zapf's Melior (1952), and although initially it conformed to a standardized modernist format, by the later 1960s it became more eclectic, combining italics and upper- and lower-case lettering to form an instantly recognizable masthead. The design of the magazines pages was equally striking. Psychedelic multicoloured typography and cartoon imagery appeared in an article of 1968, 'All You Need is Love: The Beatles in Pepperland', but by 1975 a strong 1920s 'retro' theme was evident in a fashion-led feature on 'Bathing Belles'. The fractured, sequential photography referenced film production during Hollywood's heyday and *film noir* aesthetics, and reinforced by a film-script-style typeface.[38]

Beginning in the early to mid-1960s, the supplements included profiles of pop stars, models and photographers, as well as articles on the latest fashion and design. In 1962 Mary Quant's second wholesale collection was featured in 'People of the 60s' in the *Sunday Times Colour Section*. Focusing on the idea of design as a means of communication, Ernestine Carter proposed: 'Mary Quant pioneered the taste of her generation and opened the way for others who speak the same language.'[39] Even local and regional newspapers included sections on design lifestyle, for example, the *Newcastle Journal* had its own 'Trend' section. The *Sunday Times Colour Section*, begun in 1962, was the earliest of the supplements, and from the outset it paid considerable attention to the question of design. In an early issue, its 'Design for Living' section included the feature 'A Revival for the Gothick Revival', focusing on the restored early Gothic Revival home of the sculptor Lynn Chadwick, Lypiatt Park in Gloucestershire.[40] Essentially a statement of modern, but at the same time artistic and eclectic taste, the piece invited the reader to examine the aesthetic choices made by the sculptor and his wife,

Front cover of the *Sunday Times Magazine* (15 June 1975), with eclectic masthead and a strong 'retro' fashion theme looking back to the 1920s.

Interior photo from the *Sunday Times Colour Section* (29 July 1962), showing the sculptor Lynn Chadwick with his family.

Designing Modern Britain

at the same time as offering a glimpse of their unconventional lives. Accounting for the discontinuities between the early nineteenth-century building and the modern interiors (the accompanying photographs show Chadwick's twisted metal sculptures in front of Gothic tracery windows), the reader was reminded that such open-plan spaces were just like sculptors' and painters' studios. Meanwhile, his wife Frances Chadwick, beautiful and skilled in classical French cooking, was shown in a large kitchen with Swedish cooking units, cupboards in natural pine and decorative detailing apparently copied from a reproduction Chinese Chippendale four-poster bed. The emphasis throughout was on the furnishings, fittings and designs (many by Chadwick), and how these summed up a way of life – a lifestyle – and an individual identity. The image evoked a happy family (with two small children), in a curious, but unquestionably stylish and interesting home. The 'artiness' of it was a metaphor for Chadwick's own identity, since design functioned as a social language. This particular article typified the obsession with famous people evident in the colour supplements, and it demonstrated the way in which design was used to underscore individual distinctiveness.

There were, of course, less exalted examples of design in the supplements aimed at the growing ranks of the middle class. For example, in the same year (1962) the *Sunday Times Colour Section* discussed three model kitchens for the typical family. All were equipped with a plethora of electrical appliances – cooker, washing machine, food mixer, sewing machine, tumble-dryer, refrigerator and freezer – that contributed to modern 'rational' design. Appearing in the 'Design For Living' section, these kitchens were crucial elements in the design of the home, but they represented different social identities, based on a number of themes. There was the cheaper, more accessible *Sunday Times* family kitchen that combined kitchen and living areas to produce a 16-foot-long practical space for the family. It had a dining table, sofa, storage space for children's toys and domestic items, as well as a streamlined fitted kitchen with modern appliances. Still based on the patriarchal assumption that women would be doing most of the work in the kitchen, it nevertheless offered an adaptable space 'where a mother can feed her family, do half a dozen other jobs besides cooking, and even take a nap on the sofa'.[41] The other two kitchens – an all-electric family kitchen and all-electric country kitchen – both had dining tables that aimed to make the kitchen less clinical. The former emphasized technology, but with wood-panelled walls and a stripped teak table, whilst the latter had French wall tiles copied from an eighteenth-century print to give a 'cottage' feel. The point of the three kitchens was choice. Each was meant to appeal to different aspirations and to represent distinct types of social life. The family kitchen was messy, flexible and organic; the all-electric kitchen – still a family kitchen – focused on function and practicality (the hob, worktops and refrigerator were all at hip height) and was primarily for serious cooking; and the all-electric country kitchen, with its modern gadgets and appliances, nevertheless tried to evoke a feeling for country life by referencing craftsmanship. Design in all (including the Chadwick one) was surprisingly varied and eclectic. Different styles were tolerated and traditional items of furniture incorporated, but the emphasis on technology and modern equipment remained. Equally intriguing was the position of the kitchen as the hub of the home: a large pleasant room for social interaction rather than a pokey space in which domestic labour took place. The stress on technology was meant to show that even when there was work to be done, it was considerably easier because of the abundance of modern appliances and careful design. Representations of women in the kitchen revealed not domestic drudges, but well-dressed, stylish women beaming 'her pleasure . . . in the rumblings of contentment from her man, and the delighted expressions on

Designing Modern Britain

her children's faces'.[42] Cooking at home took on new meanings, since women were expected not merely to cook, but, like Frances Chadwick, to cook very well.

Domestic ideologies reaffirming women's roles within the kitchen were re-shaped in this period, and the culture of domestic science changed too. Both boys and girls may well have been taught to cook in the new comprehensive schools established in the 1960s and '70s, but the culture of the kitchen remained focused on women. As greater emphasis was placed on cooking as a sign of familial love and as a sign of cultural capital, food became more eclectic and exotic. Elizabeth David has been singled out as a pioneer in this respect, writing on Mediterranean, French provincial and Italian cooking, but also stimulating the consumption of traditional kitchenwares imported from across Europe. Her *Mediterranean Food*, published in 1950, marked a move away from 'beige' food.[43] David too saw the kitchen as a warm, cosy family space, 'the most comforting and comfortable room in the house', in marked contrast to the pre-war modernist idea of the kitchen as a 'functional annexe' for single professionals on the move (exemplified by Lawn Road Flats in the mid-1930s). The design of these post-war kitchens symbolized the contradictions of women's roles: social change and increasing economic independence had brought new opportunities, but responsibility for the home, and particularly the stability of the family and much of the housework, still resided with women. Ironically, the emphasis on cooking well was additional to the everyday chores associated with producing regular meals: planning for the week ahead, shopping for food, washing up and clearing away. Writing in *Housewife* in 1976, Ann Oakley noted that if anything the amount of time spent on housework went up after 1945.[44]

Stainless steel and wood teapot and set, designed by David Queensberry for Viners Ltd, 1964.

Promotion and persuasion were integral to both design features and advertising in magazines. Advertising for companies such as Russell Hobbs (electric kettles), Kenwood (food mixers), Fridgidaire (refrigerators), New World (electric cookers), Ronson (hairdryers), G-Plan, Ercol and Stag (furniture), Wedgwood (ceramics) and Potterton (radiators) proliferated, and the designer's identity came increasingly to the fore: Kenneth Grange, designer of the Kenwood Chef (1961) and Ronson Rio hairdryer (1966), David Queensbury, designer for Webb Glass, Midwinter ceramics and Viners stainless steel, Robert Welch for Old Hall Tableware, Shirley Craven textiles for Hull Traders and Evelyn Brooks and Barbara Brown for Heal Fabrics, to name a few. As designers became 'stars', they lent authority and brought an aura of creativity to everyday products; as Janey Ironside, Professor of Fashion at the Royal College of Art in the 1960s found, her students were constantly in the newspapers, in fact 'the young designer had never had it so good'.[45]

As the notion that design was a language and that it communicated ideas about individual or group identities gained ground, people were encouraged to think about the meanings of design and their own consumer choices with the help of magazines and design manuals. It was, however, still possible to find the essentially modernist ethos of the COID reiterated in

Designing Modern Britain

the magazines, as in a piece in the *Sunday Times Magazine* in 1965.[46] To the question 'What's the matter with British Furniture?', came the answer that it was not innovative, did not use new materials and techniques, failed to recognize the needs of modern homes, and was too concerned with quaintness, fashion and gimmicks. Implicit here was the COID view that design was either 'good' or 'bad', and that good meant 'modern'. Consumers, however, were being exposed to an expanding range of design ideas through print and visual media, and they were becoming more attuned to 'reading' and interpreting different visual styles and ideas. Eclecticism and plurality became a feature of design in Britain from the mid- to late 1960s, and an excellent example of this was the furniture and home equipment store Habitat, which promoted design as a 'lifestyle' choice.

Habitat provided an array of household goods – furniture, lighting, textiles and kitchenware – in a shop that was more warehouse than showroom. At its opening in 1964, Terence Conran, the founder, claimed that the store provided the consumer with 'a pre-digested shopping programme'.[47] Trained as a furniture designer, Conran recognized that traditional furniture retailers provided for the general public at large, but he decided to target a particular market and to sell that market a number of coherent looks in furniture and accessories. Conran also realized that designers and retailers had to generate the 'desire' for new products, since the consumer had already satisfied the basic 'need' for them. The design of the packaging, advertising and shop interiors was crucial to this. Big, brightly coloured red-and-yellow carrier bags with Pop Art-style drawings of Habitat objects were a reminder that the shop was 'in fashion'. The stores were very simple with open-plan, quarry-tiled floors, timber shelving and discreet groupings of furniture, goods and accessories to suggest a particular 'look' or 'lifestyle'. Habitat included modernist classic furniture (by Breuer and Le Corbusier), cane furniture, ladder-back and bentwood chairs, basic stripped pine tables, tubular metal bedsteads, rustic earthenware cookware, traditional pottery (vernacular and period), ultra-simple cutlery, woven basketware, ethnic rugs and textiles, as well as plastic bathroom and kitchen equipment. An advertisement from 1971 in the *Sunday Times Magazine* offered Italian-style 'Cartel' standardized storage units, which were stackable in three colours – white, black and vermilion.[48] Starting at £4.75, these could be adapted and combined to suit all purposes, but, as the ad put it, the consumer was buying good-looking modern Italian design at cheap prices. It appealed to the young and upwardly mobile, and it allowed a certain degree of personal expression. For example, a modern low-level Habitat sofa covered with an ethnic throw might be combined with an 'English' cane chair, or an Austrian nineteenth-century bentwood chair. Equally, a reproduction

Victorian Chesterfield sofa with Art Nouveau-printed fabric covers could be set alongside a tubular steel classic. Kitchens could be rustic and country-style, 'Victorian' or modern and sleek, but they were to be equipped with diverse items: English vernacular, French country, Italian modern or Oriental. Eclecticism dominated, but it was very much of a particular ilk, and the Habitat catalogues showed the consumer how to put it all together to create an individual style that nevertheless displayed a specific lifestyle allegiance:

> The Habitat style appealed to the young and 'switched-on' because not only did it counter Scandinavian good taste and British repro bad taste, but also because '...it is an echo of their own attitude to clothes, for instance with expensive shirts worn with unpressed denims'.[49]

Habitat 'Liber' storage units combined with a Thonet-inspired bentwood chair, *Habitat Catalogue*, 1976.

From the late 1960s Habitat expanded from its original Fulham Road base in London, and with its mail-order business established, it was theoretically accessible across Britain. Stores, however, were concentrated in the south-east of England, and it was not until 1967 that a store opened in the north – in Manchester. In his role as arbiter of taste and contemporary style, Terence Conran produced a book at the height of Habitat's influence to help the consumer achieve a home that 'reflects your own personality'.[50] *The House Book*, published in 1974, was a manual outlining how to plan, design and even buy your perfect home; it listed stockists and specialists, as well as the addresses of the 20 British stores. In it Conran defined six main interior styles: farmhouse, town house, country house, Mediterranean, international and eclectic. Of these, he said:

> Each can be transplanted – but I am certainly not proposing straitjackets from which escape is impossible. Different styles from different periods can live happily together, blending into a look that reflects your own individual personality, and this is what really matters.[51]

His reasoning was that copies and interpretations of different styles enabled the consumer to develop a distinctive look, but that guidelines and constraints were needed. Questions of taste inevitably came to the fore, since the reader was advised that personal taste was the starting point, but the crucial issue was not which items of furniture to buy, but 'the way in which things come together, the background (in terms of colour and material), and the sort of accessories with which they are combined'.[52] In the

Designing Modern Britain

manner of 1930s taste manuals, Conran's lavishly illustrated colour photographs aimed to guide the consumer towards a look or a lifestyle. Plurality and diversity can be seen as central aspects of Conran's approach to design, which, by the late 1970s, was well established via the expanding Habitat stores. But, although indicative of postmodernity, Habitat's eclecticism was rarely questioning, and while Conran contributed to a shift away from the reductionism of modernism towards a visually richer, polyvalent approach to design, his agenda was in no way subversive. A consummate businessman, his strategy was to stimulate and direct consumption and to create and enhance brand loyalties.

The identification and development of visually coherent brands with distinctive 'corporate' identities took on increased significance as global markets expanded and consumer choice diversified in the post-war period. The USA led the way and designers such as Saul Bass and Paul Rand were particularly influential developing brands and logos for a variety of large multinational companies – Exxon, AT&T and IBM. Two influential corporate designers working in Britain after 1945 were F.H.K. Henrion, who established Henrion Design Associates in London, and Wally Olins, chairman of Wolf Olins, both of which were international companies.[53] Henrion, who designed, as an example, the corporate identity for the Dutch airline KLM in 1964, recognized the importance of bringing visual cohesion to the complete corporate structure, including buildings and products, advertising, stationery and packaging, uniforms and vehicles. The crucial point was that the company had to be given a clear and distinctive identity so as to communicate not just to competitors, partners and consumers, but also to articulate a corporate ethos for its own workforce. Olins took the view that a company gets the new corporate identity it deserves, and perhaps interesting in this respect was his commission in 1988 to 're-design' London's Metropolitan Police Force's internal and external image. At the time the Metropolitan Police's image was seriously tarnished by accusations of corruption, racism and its role in industrial disputes, such as the Miners' Strike of 1984–5. Dormer contends that Olins's 're-branding' bore fruit by promoting the police force in interesting ways, including a series of advertisements for newspaper, magazine and billboard hoardings that used staged social documentary photographs, hard-hitting but quite discursive texts, and some of the 'agit-prop' design ingredients of El Lissitsky.[54]

Corporate logo design by FHK Henrion for the Dutch airline, KLM, 1964.

Olins's subsequent position as a leader in the field of corporate identity design in Britain has been reinforced by the highly effective promotion and marketing of his own company. His current website presents an affable, easy-going, but essentially 'straight-talking' character who is not only good to do business with, but keen to communicate his ideas to all. This rather belies the non-benign nature of many multinational corporations and their branded products. Brands such as Coca-Cola, McDonald's, Levi's, Sony, IBM, Apple, Benetton and Nike have become global names and their logos are instantly recognized.[55] Design – in relation to brand identity, advertising, packaging and product – has been pivotal in this, but in parallel in the 1970s and '80s there was increasingly vociferous opposition to corporate power and global consumption. Within the context of radical politics in the late 1960s and theoretical debate borne (in part) of postmodernism in the 1970s and '80s, the power of the multinational came under scrutiny for ethical, ecological and political reasons.

Anti-design and Counter-cultures

An enduring icon of the late 1960s counter-culture was that of the hippy typically photographed at one of the big pop festivals held at the end of the decade: Woodstock in the USA in 1969 or the Isle of Wight in Britain in 1969 and 1970. Wearing long ethnic skirts, brightly-coloured, decorative shirts, bandanas, flowers and beads, hippies symbolized youthful rebellion, the rejection of Western materialistic society and an affinity with various political stances, including anti-war protests (Vietnam War, 1965–73), civil rights campaigns in the USA, women's liberation and the environmental movement.[56] Emerging from the late 1950s, this was nevertheless stimulated by a number of cultural events and political activities. These included the first public meeting of the Campaign for Nuclear Disarmament (CND) in February 1958 at Westminster, attended by 5,000 people, followed by the marches on Aldermaston atomic weapons establishment in Berkshire later that year; the demonstrations against US involvement in Vietnam in 1968; Allen Ginsberg's International Poetry Incarnation at the Royal Albert Hall in 1965; but perhaps more significant in terms of art and design was the student revolt about the curriculum at Hornsey College of Art in May 1968.[57] However, there was widespread protest against all types of institutions at this time, including those of higher education. Only a few months earlier, in February 1968, the Antiuniversity had been founded in London as a place for artists, activists and intellectuals from Europe, the USA and the Third World to meet without the usual divisions between staff and students and the imposition of rigidly structured courses.[58] Drawing a distinction

Designing Modern Britain

between American and European student unrest and that in Britain, Christopher Frayling noted that in Britain it 'tended to happen in the more volatile environment of the art schools', whereas in the USA and Europe it happened on university campuses, for example Berkeley and the Sorbonne.[59] Another feature of this late 1960s unrest was that it came from 'the alienated children of the comfortable bourgeoisie' and its 'inevitable elitism . . . put off many "ordinary" youngsters'.[60] There had been a significant expansion of higher education in Britain after 1945, which gathered pace in the 1960s: the number of universities had trebled; there had been a fourfold increase in students; and a significant number of young 1960s artists were working class (though, typically, male).[61] But art and design education in the 1960s was still perceived to be middle class, although interestingly this was not uniform across Britain or within all institutions. University fine art departments and London art schools had traditionally attracted middle-class students, but from the early 1970s there were a number of art and design departments established in the new polytechnics. Regional, as well as London-based, these polytechnics were often located near to centres of industry and commerce, and formed by the amalgamation of local technical colleges: for example, Staffordshire College of Technology in Stafford; the Stoke-on-Trent College of Art and the North Staffordshire College of Technology joined to form North Staffordshire Polytechnic in 1970; and Rutherford College of Technology, Newcastle College of Art and Industrial Design and the Municipal College of Commerce merged in 1969 to form Newcastle Polytechnic. These new institutions offered design as well as fine-art courses; and frequently, as at North Staffordshire Polytechnic, a curriculum linking directly to local industries – in this case ceramic design for the local pottery industry. Also they continued to attract those working-class students who had formerly gone to the local technical college; thus by the 1970s polytechnics had contributed to a distinct 'class shift' in higher education art and design provision. Student unrest in the late 1960s was, however, focused on old art schools, which were perceived by some to be bastions of class privilege. A contributory factor to the Hornsey occupation of 1968 was the effect of the Coldstream Report (1961), which attempted to raise the standard of art and design education to degree level with the introduction of academic GCE qualifications, a one-year pre-diploma followed by a three-year Diploma in Fine Art, Graphic Design, Three-dimensional Design and Fashion/Textiles. There was some concern that this academic basis (the entrance requirement was five O levels) discriminated against working-class students, and that the over-emphasis on the exam requirement (stemming from the desire of those institutions offering the Dip. AD for degree equivalence with universities) underplayed vocational work.[62] Another target of the protesters was 'the separation of departments into Fine Art, Graphics, Three-Dimensional Design, and Fashion/Textiles [which was] was anathema to young people interested in informal networks and flexible systems'.[63] At the same time artistic hierarchies privileging fine art over some aspects of design were clearly apparent, and these were reinforced by gender stereotypes too, as Barbara Hulanicki, the founder of Biba, described. Enrolling in a life class as a student at Brighton School of Art, she found herself summarily dismissed by the tutor as 'another fashion one'.[64]

Intrinsic to the wider political questioning of the late 1960s was a developing critique of consumerism. Herbert Marcuse, a member of the Frankfurt School, was particularly influential for 1960s radicals, particularly his challenge to the idea of material progress and technical advance found at the core of modernism. Writing from a position on the left, Marcuse was critical of organized political systems, believing instead in the effectiveness of marginal groups. A number of crucial books were published that pointed the finger at large multinational companies. Perhaps most important in relation to design were the writings of Vance Packard (the best known was *The Hidden Persuaders,* 1957). Also influential was the work of Ralph Nader, particularly his *Unsafe at Any Speed: The Designed-In Dangers of the American Automobile* (1966), which took the US automobile industry to task for its poor safety record.[65] Parallel to this was a growing recognition of the need for consumer protection, and in Britain, according to Woodham, 'the government undertook a more active role with the establishment of a national Consumer Council in 1963 and the passing of the Trades Description Act in 1968'.[66] Also related to the debate about big business were issues of ecology and ethical trading as organizations such as Oxfam began marketing handicrafts from the developing world in 1964 in order to give small-scale producers fair prices, training, advice and funding.[67] Imported baskets, textiles, rugs, ceramics and other indigenous crafts could be found on the high street in Oxfam and other similar outlets. Ecological issues also came to the fore in Third World countries and in Britain. In the latter, these were highly visible on British TV screens when the *Torrey Canyon* oil tanker came aground in the spring of 1967 in the English Channel. The first of the big supertankers, she broke up carrying a cargo of 120,000 tons of oil that washed up on the south coast of England, killing the sea biology of the region.[68] The Body Shop, set up by Anita Roddick in Brighton in 1976 to sell cosmetics, adopted an ecological stance with its biodegradable plastics and its related practice, whereby plastic containers could be returned to the shop for refilling. Buying goods from The Body Shop involved 'not simply buying a product but buying into a whole value system'.[69] The packaging was minimal – simple plastic bottles with basic, clear labels, and a minimum of advertising and promotion. The Body Shop website in 2006 effectively sums up ambitions first articulated in the late 1960s and early 1970s: 'against animal testing'; 'support community trade'; 'activate self-esteem'; 'defend human rights'; and 'protect the planet'.[70] But ironically, in this same year, 2006, the French cosmetics manufacturer L'Oréal bought The Body Shop for £625 million. The company had succeeded with a distinctive formula, but its global expansion (more than 2,000 stores in 53 countries at the point of takeover) meant that it was far removed from the small independently-minded shop established 40 years ago. Indeed, setting aside L'Oréal's poor reputation in relation to animal testing, many of its supporters believed that The Body Shop (described as a 'Small Giant' by Roddick) had left its ethical principles behind when it became a global organization – irrespective of Anita Roddick's continued commitment to good causes such as anti-sweatshop legislation, human rights campaigns and pro-women initiatives.[71]

Ethical, consumerist and ecological questions stimulated design groups, retail organizations, educators and designers. Radical architectural groups such as Archigram in Britain had already challenged aspects of architectural practice and the profession since the mid-1960s. Radical forms of design practice were also a feature of design in the mid-1960s with inflatable, cardboard and disposable furniture, but a new ingredient in the late 1960s was the developing critique of objects as markers of status and conspicuous consumption, combined with a growing anxiety about the waste of the world's limited natural resources. Awareness of this was growing across the developed world, and it had a profound effect on some designers. For example, anti-design initiatives in Italy stimulated a number of alternative strategies that brought into question the processes and rationale of designing specific objects for personal, business and industrial consumption. Described by Penny Sparke as the 'crisis of the object', such debates were 'less significant for the objects [they] inspired than for the ideas [they] stimulated'.[72] The exhibition and accompanying catalogue *Italy: The New Domestic Landscape* at the Museum of Modern Art in New York in 1972 gave these ideas institutional legitimacy.[73] To a certain extent the designs emerging from groups such as Superstudio and Archizoom, which questioned the never-ending production and consumption of individual design objects by a rich, indulgent society, echoed some of the ideas of the American architect and thinker Richard Buckminster Fuller. His geodesic domes that could be constructed from industrial waste, for example, struck a cord with those aiming to curtail the production and range of design goods, and gained an added poignancy as Western economies faced the prospect of high oil prices and recession after 1972–3.[74]

An altogether more concrete example of radical design practice and designer roles arrived in Britain in 1976 with the 'Lucas Plan'. This was an alternative proposal to the rationalization plans put forward by the senior management of Lucas Aerospace, one of Europe's largest designers and manufacturers of aircraft and equipment. With 18,000 workers spread over 15 factories across Britain, the management plans would result in the closure of some factories and the loss of 20 per cent of jobs. In response, an alternative corporate plan was developed involving all workers (secretaries as well as those on the production line), which analysed skills, machinery and equipment in the company's factories before proposing that the company widen its product base by shifting from military production towards more socially useful goods. The latter would include a considerable increase in the production of medical equipment, such as kidney dialysis machines, as well as new developments, such as a 'design for disabled' unit, energy-saving heat pumps, solar cells and a flexible power pack easily adaptable to the needs of under-developed countries.[75] The Lucas Plan, which was ultimately rejected by the company, highlighted ethical issues in relation to design, but it also proposed specific design solutions in order to address them. It questioned the real usefulness of military goods for society, and called for a fuller consideration of the resources – financial, human and natural – required in the production of these goods.

Such practical proposals found a theoretical rationale in the writings of Victor Papanek and Gui Bonsiepe, and through the organization of important conferences in London, such as the International Council of Societies of Industrial Design (ICSID) in 1969 on 'Design, Society and the Future', and in 1976 on 'Design For Need' (the latter at the Royal College of Art).[76] Papanek's influential book *Design in the Real World: Human Ecology and Social Change*, first published in English in 1971, made an enormous contribution. With shocking directness, his first sentence began: 'There are professions more harmful than industrial design, but only a very few of them'.[77] To a generation brought up to believe that the designer acted for the greater social good by contributing to the design of a modern technologically progressive society, Papanek's invective was hard to swallow. He demanded that designers learn to be more responsible, and in the preface to the second paperback edition in 1985 highlighted the dubious trading practices of the so-called developed world in the developing world:

> The road between the rich nations of the North and the poor southern half of the globe is a two-way street. It is reassuring to understand that designers in the Third World can solve their own problems free from interference by 'experts' imported for two weeks.[78]

Designing Modern Britain

Traidcraft, established in Gateshead, England, in 1979, was one of a number of organizations in Britain that developed a response to such ideas. A Christian enterprise, with a substantial retailing arm, it sold clothes, food and crafts via a mail order, and latterly an online, catalogue in the predominantly rich 'Northern' hemisphere countries. Its main aims were to fight poverty by engaging directly in trade so as to develop the skills and market access of poor producers; to enable those in poor countries to develop and use their skills; and to encourage 'responsible stewardship of the created environment, giving people access to resources, a share in decisions about them, and responsibility for their use'.[79] Its retailing rationale was to provide a market for goods such as crafts that employed poor people in developing countries. Labour-intensive crafts and clothing were favoured because these were often highly finished at source, thereby giving maximum value at the producer end.[80]

Ethical trading and design probably had more of an impact at a theoretical level than a practical one. However, as design goods from the Third World gained in popularity and availability, they contributed to changes in taste, particularly in the 1970s, but also subsequently. The burgeoning fashion for clothing and goods from African and South American countries and from the Indian subcontinent gathered momentum in the late 1960s and '70s, and although ostensibly signifying allegiance to the political agenda of the counter-culture, by the mid-1970s 'ethnic' was but one of a plethora of styles. Ethnic goods and traditional Asian forms of dress were also evident on Britain's high streets, particularly from 1948 to 1969, when workers from India and Pakistan migrated in large numbers to Britain. From the early 1970s small independent fashion boutiques, ethnic importers and market stalls across the country sold an eclectic mix of 'ethnic' craft goods: oriental jewellery and ornaments, incense sticks, richly coloured and patterned clothing, leather goods, scarves and accessories. These outlets and styles were often to be found in marginal areas of the city: Camden Lock in north London was an example. Of the street markets, Camden Market was (and still is) the most well known and extensive, and then as now it offered kaftans and embroidered jackets, cheesecloth shirts and skirts, tooled leather sandals and bags, beads and scarves, as well as antique clothes and records, crafts, and ethnic and health food. In the early 1970s it was a depressed area of decaying Victorian houses and empty and boarded-up shops, but has since become 'a target for those employed in the media and higher education-sectors who spearheaded a familiar process of gentrification in NW1'.[81] An example of the independent boutique was Boodle-Am in Leeds, established in 1969, which sold ethnic goods as well as retro clothing and accessories. It was part of a clutch of alternative shops and stalls in Leeds selling second-hand clothes, ethnic goods and crafts. It was

located in the Victorian County Arcade in Leeds, which, like the Handiside Arcade in Newcastle upon Tyne, was rundown and on the fringes of the central modernized shopping area. In Newcastle, the Handiside Arcade (built 1903) represented an enclave of alternative shopping with its antique and junk shops, second-hand, ethnic and retro clothing shops, record and hi fi shops. In Leeds and Newcastle, along with other large British cities, low-rent space was available for the development of an alternative, small-scale retail culture, which, although driven by commercial needs, was to varying degrees a response to the counter-culture of the late 1960s–early 1970s. Significantly, this was neither 'modern', nor 'English', but instead highlighted the cultural diversity of post-war Britain.

Folklore Waistcoat, from Traidcraft Mail Order company, mid-1980s.

Market-stall holder selling 'ethnic'-inspired goods at Camden Market, Camden Lock, north London, 2006.

Past Times and Poverty Dressing

By the late 1960s and early 1970s disillusionment with consumerism and modernity permeated aspects of design in Britain. In part prompted by social, economic and political factors, there was a parallel resurgence of interest in the past, particularly from before 1945. Raphael Samuel, arguing that 'the nostalgia industry' was a phenomenon of this period, also saw it as a reaction to the widespread modernization that had seen the 'old' and 'historic' systematically demolished and rejected.[82] Conservation groups such as

Designing Modern Britain

the Victorian Society, founded in 1958, were involved in landmark campaigns to protect nineteenth-century buildings, but there was an interest in the past at a popular level too. Hollywood played a part with films such as *Bonnie and Clyde* (1967), *Chinatown* and *The Great Gatsby* (both 1974), fuelling an interest in 1920s and '30s fashion and design. Art galleries and museums staged large exhibitions that examined these periods: in 1971 Cecil Beaton organized an exhibition at the Victoria and Albert Museum in London of his 'dream wardrobe', which included clothes by Poiret, Schiaparelli, Lucille and Fortuny.[83] In 1975 the exhibition *Fashion, 1900–1939* started at the V&A before moving to the Royal Scottish Museum in Edinburgh, with essays in an accompanying catalogue by, among others, Madge Garland, Madeleine Ginsburg and Martin Battersby. Battersby's books, *The Decorative Twenties* (1969) and *The Decorative Thirties* (1971), contributed to the growing taste for Art Deco.[84] In 1979 the Hayward Gallery in London staged *Thirties: Art and Design before the War*, which introduced people to the architecture, art and design of the 1930s. Meanwhile, the popular press and magazines rediscovered the work of relatively unknown artistic and cultural figures from before the Second World War. A series entitled 'The Master Builders' in 1971 in the *Sunday Times Magazine* had already included articles on the Wood partnership responsible for building the Royal Crescent in Bath. But indicative of the new vogue for art and design from the late nineteenth and early twentieth centuries was number five in the series – a piece on the work of the American Arts and Crafts architects Charles and Henry Green by Reyner Banham. An attraction of the interwar period was that some of the participants were still alive; in 1975 the *Sunday Times Magazine* ran a feature on the designs of Eileen Gray. Described as the 'quiet precursor of "art deco"', she was interviewed in her home in Paris aged 96. The stylish and exotic photographs and examples of her work found in the apartment on the rue Bonaparte pointed to a hitherto unexplored history of complexity and contradiction in architecture and design.[85] Gray was, in fact, typical of several designers from this era (significantly many of them women) whose work had been bypassed by those interested in a narrowly defined modernism, and as a result, they had been forgotten. Yet the evidence pointed to an alternative modernity, abstract and avant-garde, but also decorative, visually complex and luxurious. It also contributed to the growing feminist interest in those 'hidden from history'.[86]

For the proponents of postmodernism such as Charles Jencks, the traditional, the alternative, the decorative and the historical provided the preconditions for 'radical eclecticism and multiple-coding'. Complexity and contradiction were welcomed, the result of a combination of historical styles, modern technologies and a self-referential, tongue-in-cheek knowingness. Others offered a more critical perspective of the cultural eclecticism of the late 1960s and '70s. Frederic Jameson, for example, highlighted the symbiotic relationship between postmodernism and what he described as 'this new moment of late, consumer or multinational capitalism'.[87] Consumption was critical to this process, and he dated it after 1945, when

> A new kind of society began to emerge (variously described as postin-dustrial, multinational capitalism, consumer society, mass society and so forth). New types of consumption; planned obsolescence; an ever more rapid rhythm of fashion and styling changes; the penetration of advertising, television and the media generally to a hitherto unparal-leled degree throughout society; the replacement of the old tension between city and country, center and province, by the suburb and by universal standardization; the growth of the great networks of super-highways and the arrival of automobile culture – these are some of the features which seem to mark a radical break with that older pre-war society in which high modernism was still an underground force.[88]

Setting aside the obvious comparisons with early twentieth-century perceptions of modernity, to Jameson a crucial feature of this late twentieth-century perspective was the endless reworking of various styles. Drawing a distinction between pastiche and parody, he suggested that eclecticism involved both parody and pastiche. Whereas parody produced an imitation that mocked the original, pastiche was parody without the knowingness: 'It is a neutral practice of such mimicry, without parody's ulterior motive, with-out the satirical impulse.'[89] In contrast, Jencks's view was that this added richness and diversity to architectural language after a sustained period of the 'less is more' philosophy of high modernism. Both Jameson's and Jencks's ideas are important in any discussion that aims to understand the interest in period styles, alternative cultures and indigenous forms evident in design towards the end of the 1960s and early 1970s. As Angela McRobbie argued in her pioneering work on second-hand clothes and rag markets, however, postmodernism offers just one way of approaching the eclecticism of much design in this period. She located the taste for second-hand clothes firmly within the economics of the early 1970s, and noted the importance of entrepreneurial young women in this particular subculture.[90] Furthermore, a company such as Habitat produced neither straightforward pastiche nor self-referential parody, even though it looked beyond modernist design solu-tions to provide the consumer with diversity in design and lifestyles. It set about, as Dick Hebdige has argued, to promote 'syntax selling'.[91] In seeking to establish the specific social foundations of taste,

Terence Conran's selection of designs for Habitat are being used to shape and frame what Pierre Bourdieu calls the 'habitus' – the internalised system of socially structured, class specific gestures, tastes, aspirations, dispositions which can dictate everything from an individual's 'body hexis' to her/his educational performance, speech, dress and perception of life opportunities.[92]

A precursor of syntax selling, or 'the cultivated habitus', was the fashion boutique-cum-store Biba.

Biba, a mail-order company set up by Barbara Hulanicki and her husband, opened as a small boutique in Abingdon Road in 1964. Because of its phenomenal success, it moved to larger premises in Kensington Church Street, and then to Kensington High Street, before finally closing as 'Big Biba' in Derry and Toms' Art Deco department store in 1975.[93] It began as a rather ad hoc enterprise based on the 'fashion eye' of Hulanicki, whose first Biba design was a pink gingham shift dress that attracted the eye of the *Daily Mirror*'s fashion editor. She was inspired at this time by popular culture, particularly the gamine-like Audrey Hepburn and the youthful Brigitte Bardot (apparently wearing gingham at her wedding to the singer Sacha Distel).[94] It quickly metamorphosed into a boutique selling polka-dot and striped shift dresses and skirts, as well as hats and jewellery from a small dilapidated old chemist's shop, to a substantial business selling opulent luxurious suits, dresses, make-up and all manner of accessories. The interior of the first boutique established an approach evident in subsequent design. It was dimly lit with dark blue walls and it had a late-Victorian feel, with screens providing a place to change. By 1967, following its move to Kensington Church Street, it took a decided 'retro' turn, drawing on Art Nouveau in particular, which was revived in the 1960s via graphic design and posters, and inspired by, but also contributing to, a renewed interest in the work of designers such as Aubrey Beardsley and Alphonse Mucha. The boutique's main window was painted black, except for John McConnell's Art Nouveau-inspired Biba logo in gold, and the interior, incorporating the original wooden shop-fittings, had a dark red changing room referencing a 'Bordello'.[95] Graphic design and packaging were particularly important, and the first Biba logo in black and gold summed up the *fin-de-siècle* mood with its Celtic interlacing swirls. McConnell designed for Biba between 1965 and 1972, producing the catalogue and packaging, and in 1969 redrawing the logo, using Art Deco motifs as inspiration. He later joined Pentagram, formed in 1972 by Alan Fletcher, Colin Forbes, Theo Crosby, Kenneth Grange and Mervyn Kurlansky. Pentagram, which developed in part in response to 1960s pop culture, deployed wit alongside visually clever design

elements, such as the mailed brown parcel tied with string that became the cover of *Graphis* magazine and the Pirelli slippers bus poster, both from 1965. These were quirky and eclectic referencing distinctive 'British' icons and institutions – the double-decker bus and the Royal Mail – but pointing beyond national boundaries in, for example, the standardized, repetitive dot design for Reuters news agency in 1968.

Distinctive graphics contributed to the visual diversity of late 1960s and '70s design, and certainly at Biba, McConnell's work helped to define the 'Biba' look. Signalling a clear difference from typical mail-order catalogues, his long, slim format Biba catalogue with dual-tone and unusual photography was a precursor of the Next Directory in defining 'lifestyle'.[96] But, as with the cover of *Graphis*, there was a subversive, oppositional stance element to these designs. Biba clothes provided 'elegance with attitude' and offered '"grown up" clothes but . . . on youth's terms'.[97] Popular fabrics included satin, crêpe, chiffon and velvet in rich muted colours: blackish brown, prune, dusty pink, duck-egg blue, bilberry, rust and mulberry. The shapes included wide trousers and floaty dresses with sleeves tight to the elbows then flaring dramatically. Biba 'femininity' was exaggerated, vampish and playful: 'little girls dressed up as women' with faces that had a 'stoned, trance-like quality'.[98] Hulanicki, trained as an illustrator, had considerable knowledge of fashionable styles (regularly covering the Paris collections as an illustrator), but she was attuned and contributed to a significant change in taste towards the end of the 1960s. As a consequence Biba can be seen to exemplify a pivotal shift in design in Britain from 'the sophisticated minimalism' of the mid-1960s to the 'sulky self-awareness' of the late 1960s and early 1970s.[99] As Hulanicki put it: 'as time went by my

Catalogue for Biba Mail Order, 1969, with John McConnell logo.

Catalogue for Biba Mail Order, 1968.

Designing Modern Britain

Biba girl became more dreamy and untouchable. Her long straight hair turned into a halo of golden ringlets, her cheeks were hollowed by brown powder, and her lips stained with sepia lipstick.'[100] Hulanicki described this image as mysterious and ambiguous, and recalling Terence Conran's 'syntax selling' at Habitat, she observed: 'We gave them basics that they could then interpret in their own way. I could sit in the shop for hours watching customers who had previously bought things and now were wearing them with a completely new slant.'[101]

Reading Hulanicki's autobiography it is difficult not to conclude that her approach was influenced by her unusual background, particularly the authoritarian, gin-drinking 'auntie' who wore couture clothes –'living by 1930s fashion rules' – and exercised financial control over the extended family.[102] Eager to escape this, fashion provided some measure of common ground between them, but perhaps informed by her own experience, Hulanicki's notion of the 'ideal' Biba customer, 'looked sweet but was as hard as nails. She did what she felt at that moment and had no mum to influence her judgement.'[103] There was certainly a measure of transgression and defiance in the later Biba 'look', particularly in the referencing of 1920s and '30s Art Deco, Hollywood glamour and *fin-de-siècle* Art Nouveau. As an early Biba fan put it, 'anti-establishment attitude was the touchstone by which we judged everything in 1968, and there was no doubt at all that by our criteria the Biba mail-order catalogue scored very highly: they definitely had attitude'.[104] Hulanicki's reference points were unquestionably the recent past: William Morris prints decorated the shops; Celtic and Art Nouveau motifs were used for labels, packaging and posters; and antique wardrobes, hat stands, chaise longues and potted palms were props in the increasingly theatrical shop interiors. But this was put together in an eclectic mix, a complex 'multiple-coding' not just to create a retro-chic style, but also a different approach and philosophy.

Somewhat naive to begin with, Hulanicki's attitude to the fashion business was ultimately challenging. She liked texture, pattern, decoration and richness; she collected second-hand, antique and reproduction items of clothing and design; and she seemed to recognize that the female consumer was not a passive dupe. There was a subversive quality to her sense of taste and style; it was opulent, messy and inconsistent, and she made these positive qualities. She commented on how the serious end of the fashion trade had little interest in her, but in the popular press she had a huge following. She was certainly critical of the established rag trade, which she claimed seemed to assume that in order to sell, clothes had to be bland: 'Designers at that period seemed to feel they should dictate to women. I couldn't see how they knew what women wanted – those designers were usually men.

The young wanted to be led but not dictated to.'[105] Biba's studied stylistic decadence, nostalgia for the past and unconventional business methods responded to, but also reinforced, late 1960s counter-cultures and the enthusiasm for designed 'retro' styles, as well as second-hand clothes.

Second-hand clothes were particularly appealing for their shock factor; indeed, they were 'a terrifying reminder of the stigma of poverty, the shame of ill-fitting clothing, and the fear of disease through infestation'.[106] But importantly, in the late 1960s and '70s these were carefully selected 'old' things that were combined with new items, thus allowing both the display of 'cultural capital' and a subversive engagement with fashion. Second-hand style was dependent on an eclectic mix: second-hand clothes and acces-sories, antique clothing (from different eras), new items and traditional ones (fair-isle cardigans and hand-made shawls, for example). Rarely did anyone dress in a coherent 'set' of second-hand clothes; the aim was to sub-vert and confuse the language of clothes by deploying and exploiting fashion knowledge. Interest in the old also implied a refusal of the modern and, by implication, the artificial, urban and man-made, and there was a concurrent enthusiasm for authenticity and the rural; 'real' crocheted shawls, hand-made footwear, traditional tweeds and hand-made lace were revived. Typically, fashion designers and retailers exploited all this. Laura Ashley, who began producing printed textiles in the early 1950s, started sell-ing women's clothes in the mid-1960s using simple fresh country-style cotton prints with small repeat patterns of flowers with Edwardian and Victorian design details, such as smocking and leg-of-mutton sleeves, and decorated with pin-tucking, lace and embroidery. Other designers such as Margaret Howell, who had trained at Goldsmiths College of Art, launched her own range of very 'English' clothes in 1972. She specialized in using wool tweeds combined with brogues, capes and traditional suits in rich natural hues, but with a modern quirky cut. As we will see in chapter Six, the enthu-siasm for 'heritage' was consolidated into mainstream design (graphics, fashion, furniture, textiles) and architecture in the 1980s and '90s, but there were already one or two moments of resistance when the desire to under-mine this groundswell of nostalgia for times past was ascendant – probably most provocatively with Punk.

Second-hand clothes provided part of the ancestry of Punk,[107] but it also represented an 'open identification with black British and West Indian culture'.[108] Rasta language, hair, rhythms and national colours coalesced with class to mark the boundary with the glam rock of David Bowie and Roxy Music, although the sub-text of glam rock – sexual and gender ambi-guity – certainly influenced Punk female style during its brief period of ascendancy in 1976–7. Arguing that Punk's intentionally 'gutter-snipe' rhetoric

Women's fashion from
Laura Ashley, late 1970s.

Interior design from Laura
Ashley Ltd, mid-1980s.

was to set it apart from the attitudes and values of the previous generation of rock 'stars', Hebdige interpreted the Punk aesthetic 'in part as a white "translation" of black ethnicity', and noted that Punk's fascination with reggae stemmed from its identification with something perceived as a threat to mainstream British culture.[109] There was also a dynamic relationship between black British sub-cultures and African-American ones, and in this music was vital, particularly the powerful role of reggae and soul music with their political connections to Rastafarianism, civil rights and Black Power. 'Young, gifted and black', as Bob (Andy) and Marcia (Griffiths) sang in their Top Ten British hit in 1970, was an evocation of awakening pride as afros attained 'cool' status, whereas for those following Bob Marley 'the tracksuit became de rigueur'.[110] Black women singers projected powerful black visual identities, including soul singers such as Aretha Franklin, but also Marcia Griffiths, one of the I-Threes, backing singers for Bob Marley. The status of women in Rastafarianism was not, however, unproblematic: 'out of obedience to the doctrine, Rasta Queens covered their locks at all times in public, swathed themselves in draped fabric and wore humble sandals.'[111] In contrast, the embrace of androgyny, an angry femininity and the denial of romantic love became part of Punk's sub-cultural language and it offered a means by which 'the critical ambiguities of women's presentation of themselves' could be explored.[112] Referencing pornography, these 'clothes had to be worn with both aggression and irony, in the knowledge of what they signified'.[113] Determinedly DIY, devoid of 'taste' and contradictory, Punk style used fabrics and patterns that were tacky and gaudy. Modern materials such as Lurex, fake leopard skin and PVC were combined with fluorescent green, shocking pink, vivid orange and, of course, black. Artificiality characterized make-up and hair, which was extreme and abstract; unfinished and torn seams were turned inside out; old ties and bits of school uniform were added to skimpy and provocatively screen-printed T-shirts; torn trousers and skirts sculpted with safety pins, multiple zips, chains and stilettos, were 'the illicit iconography of sexual fetishism . . . used to predictable effect'.[114] Neither looking back to some mythical, authentic 'English' past, nor forward to a technocratic, progressive future, the meanings of Punk were a subversive response and contribution to the present; a visual and material product of the ambiguities of progress discernible in 1970s Britain. The DIY aesthetic evident in second-hand clothes and Punk fashion could be seen in magazine design since the 1960s, thanks in part to the influence of Pop Art: torn-out images and rough assemblages of text and collage. This anti-aesthetic, seen in the design of the satirical magazine *Private Eye*, influenced magazine design from the mid-1970s, such as Punk fanzines (*Sniffin Glue* and *Ripped and Torn*), but it also contributed to the visual style of magazines as diverse as the feminist monthly *Spare Rib* (designed by Katy Hepburn and Sally Doust) from 1972; *i-D* (art director Terry Jones and designers including Moira Bogue and Stephen Male) from 1980; and *The Face*, launched in 1980 (designed by Neville Brody from 1982 to 1986). Close links to the British pop scene, and particularly to album covers designed by, for example, Jamie Reid (the Sex Pistols), Barney Bubbles (Elvis Costello) and Malcolm Garrett (the Buzzcocks), enabled graphic designers to avoid the overworked, slick imagery of mainstream magazine and graphic design to produce striking visual styles.[115]

A dominant characteristic of design in Britain in the late 1960s and '70s was diversity. This was partly rooted in the processes of critical questioning widespread throughout Western societies from the late 1960s, but it also stemmed from the rejection of orthodox or 'high' modernism promoted nationally and internationally in the 1950s and '60s. In such a context an interest in historical styles (even those from the previously maligned Victorian period) and an enthusiasm for non-Western cultures, particularly from the Indian subcontinent, developed, but these combined with the plethora of cultures emerging within Britain's own black communities. Alongside this, debates about the ethics of consumerism, global trading and ecology also contributed to design in Britain. As the post-war economy became more competitive and the marketplace global, there was an overrid-

The first issue of *Spare Rib* magazine cover designed by Katy Hepburn and Sally Doust, July 1972.

Designing Modern Britain

no swim swam!

B'BOYS AT THE KREMLIN!
GOOD TIMES BAD TIMES!
THE SMITHS! Morrissey

hiphop to Moscow
affluence in working Britain
interview

PLUS

The Cramps· Cristina
Philip Sallon· Tom Verlaine
Daryl Hannah· Videopop
Honda Imps

James Truman: Why do painters have perfect teeth?

Cover design of *The Face*, July 1984. Neville Brody provided art direction from 1982 to 1986.

ing need to sell additional goods to a consumer who had acquired the skills of discernment. Advertisers, manufacturers and retailers became adept at promoting products on the basis of their wider meanings. These were constituted within a matrix of social, political and economic change in post-war Britain. Greater opportunities for women and working-class people, combined with cultural diversity borne of immigration by West Indian and Asian Britons settling in Britain's large towns and cities, brought a plurality that undercut white, middle-class, male value systems. Parallel to this were various theoretical debates informed by Marxism, feminism, post-structuralism and postmodernism, challenging the position of the white male subject and his vision of technological and social progress. To a large extent the dominance of 'orthodox' or high modernism had been a product of this 'subject', but as the challenge to modernism's cultural authority gathered pace, it became apparent that these universal solutions offered a very limited repertoire of ideas, especially in an age of mass consumption.

6

'I Shop Therefore I Am'[1]: Design since the '80s

Two years after Margaret Thatcher opened the new Terence Conran-funded Design Museum in London in 1989, Robert Hewison drew a rather apt analogy between it and the original South Kensington Museum founded in 1852: 'Lurking within the converted warehouse at Butler's Wharf is the ghost of Henry Cole's Museum of Manufactures, evicted from South Kensington when the V&A took . . . its "antiquarian turn".'[2] The South Kensington Museum (later the Victoria and Albert Museum), founded with examples of design from the Great Exhibition at the Crystal Palace in 1851, was a means to instruct manufacturers and the public in the rules of good design and taste. Cole, its first director, sought a close relationship between art and industry, but for Hewison, who believed that museums were emblems of society at large, it was all too revealing that the Design Museum's first exhibition should be entitled *Commerce and Culture*.[3] In his view, the Design Museum had overemphasized the commercial aspect and lost sight of the cultural. In the catalogue accompanying the exhibition, the museum's first director suggested: 'Shops and museums have a great deal in common. Urban, predominantly middle class, dedicated to exhibition, committed to consumption, either of images, ideas, or goods.'[4] The Design Museum was not the only museum to attract Hewison's critical eye. The Victoria and Albert Museum's 'antiquarian turn' was reforming the history of design into just another marketable commodity: 'Laura Ashley and the V&A, Habitat and the Design Museum, will simply be competitors in the same vast cultural market-place', he wrote.[5] Museums and their exhibits were being commodified within a burgeoning enterprise culture, which affected definitions of taste and good design, which, in the case of the Design Museum, included the Conran-owned retail stores – Heals, Habitat, Mothercare and British Home Store – as well as the restaurant, Bibendum.

Air Chair (1999). Produced by Magis, designed by Jasper Morrison.

In the predominant language of the right, however, an injection of 'enterprise' offered the solution to Britain's economic stagnation, and design had a crucial part to play in this by stimulating innovation and experimentation, the like of which had not been seen since the Great Exhibition of 1851.

The politics of the right dominated the final 20 years of the twentieth century in Britain and with this came a return to so-called Victorian values, particularly self-reliance and individual enterprise. A crucial element in achieving this was the 're-presentation of national identity for a post-imperial age'.[6] At the core was an attempt to rekindle a Victorian spirit of entrepreneurial activity, while at the same time evoking specific notions of 'Britishness' and 'Englishness'; this required the unpicking of structures set in place after 1945 for social equality and justice, and the parallel commitment to full employment and the Welfare State. With no such thing as society,[7] identity was defined not by class, gender, race, ethnicity, generation or sexuality, but in the marketplace and through consumption practices. Identities were evoked and represented via the commodification of particular ideas of 'Englishness' and 'Britishness', invoking a highly selective reading of the past that stressed the virtues of free-market capitalism in the eighteenth and nineteenth centuries. These processes of commodification involved the marshalling of 'heritage' in design and material culture, and, as we will see, the past was not so much a 'foreign country', as a well-worn reference book, falling open at certain pages due to over-use. But what of those visions of a more cohesive, egalitarian society conceived by modernists at the start of the twentieth century and proposed by those on the left at its close? In an attempt to wrest the agenda from those on the right, Corner and Harvey argued: 'Far from our future being *given* in our past, the past suggests a variety of possible and different sources of action, including economic innovation that refuses the principles of the "free" market.'[8]

Design in late twentieth-century Britain was formed by, but also contributed to, a matrix of ideas, values and attitudes. There was a continuing engagement with broader ethical, political and social questions that were manifest in relation to ecology, feminism and economic regeneration. A mixture of enthusiasm and anxiety characterized the reworking of essentially modernist ideas, particularly regarding technology and progress; this was evident superficially in relation to building technologies and the celebration of late modernism, but there was also deeper questioning of the plethora of information technologies and their wider impact on people's everyday lives. An ongoing engagement with the politics of identity and consumption formed part of a larger debate about the impact of consumerism. These took place within the context of organized politics (from both the right and the left), but they also emerged in relation to specific issues and/or identities. Questions relating to femininity provoked interrogation of what it meant to be masculine, particularly within the context of deindustrialization and mass unemployment in the 1980s. Alongside this, gay identities and those based on race (including 'whiteness'), ethnicity and religion came to the fore, and as the post-war baby boomers moved into their 40s and 50s, questions of ageing and generation came into view. Permeating all this were the increasingly complex issues of geography and identity, particularly international, national and regional, but also urban and rural. These were articulated within the framework of Scottish and Welsh devolution and the prospect of regional assemblies, but also in the bigger European and global arenas.

This chapter explores the relationships between late twentieth-century reinventions of modernism and postmodernism(s) and shifting perceptions of identity by focusing on several themes in design: fashion and lifestyle marketing, the status of the designer, nostalgia and design, and design and enterprise cultures. The focus on the individual during almost 20 years of Conservative government (1979–97 with three election wins – 1979, 1983 and 1987 – under Margaret Thatcher) had a very specific effect on design by privileging 'the designer' and design-led companies. Design languages articulated social distinctions, and particular categories of goods, retailers and designer 'names' became crucial markers of cultural capital. The collapse of British manufacturing industry and the subsequent shift to a service economy put the emphasis on the consumer, and as a consequence, marketing, retailing and advertising became even more critical. Shopping centres, designer outlets and latterly internet shopping have transformed the marketplace, and huge out-of-town retail complexes have undermined the viability of established shopping centres in towns and cities. As design in Britain fragmented, a variety of lifestyles were marketed, and theoretically all consumers catered for. Design in the 1980s and '90s was then typically postmodern: fragmented, incoherent and eclectic, but with some potential for radical questioning. Some designers engaged in contestation, especially young fashion and graphic designers (including Vivienne Westwood, John Galliano, Katharine Hamnett and Neville Brody). Their designs represented an anti-aesthetic, but potentially their 'designer' status undermined any serious political comment they might try to make.

Global structures of production, promotion and consumption led to the 'internationalization' of design (consider the product ranges and markets of, for example, mobile phones and computers), but, ironically, at the same time the reworking of visual styles drawn from periods of British 'imperial' power abounded. The past was reinvented via major exhibitions in museums and art galleries, in nostalgic advertising campaigns, in television programmes and in blockbuster films. The heritage industry and popular nostalgia about the past provided powerful images that were taken up by all types of designers. History programmes joined cookery, gardening and interior design as popular TV, and designers proved adept at recycling the Georgians, Victorians and Edwardians and the 1940s as part of their never-ending search for something new. This was typified in the market for high fashion – witness collections by Paul Smith and Vivienne Westwood in the 1980s and '90s – but equally high-street fashion (for example, Jigsaw) also traded on aspects of tradition as tweeds and kilts, knitted Arran and fair-isles, riding boots and brogues, hand-made lace and crochet were incorporated into new ranges. In contrast, IKEA's advertising campaign of 1996, 'Chuck out your chintz', showed a disregard for cosy middle-class 'Englishness' in favour of a more abstract and eclectic design language that appealed to an international market. The nature of design also changed, with essentially modernist ideas informed by technical innovations, creating new working practices, processes and materials. A large group of British designers – mainly trained in Britain's art colleges and universities – contributed to this, producing highly modern, inventive and stylish designs for British and international companies.

Wallpaper* magazine, launch issue, London 1996. Dedicated to 'international design interiors and lifestyle', the magazine also celebrates fashion and travel, and is targeted at a young, affluent, urban readership.

Re-making Britain

Ironically, in 1987, the year that Margaret Thatcher led the Conservatives to a third consecutive victory by questioning, among other things, the main opposition party's ability to defend Britain, Robert Hewison published his hugely entertaining and highly critical *The Heritage Industry*, revealingly subtitled *Britain in a Climate of Decline*.[9] Doubting the 'real' figures for production and employment, he proposed that the country was 'gripped by the perception that it is in decline'.[10] Subsequent historians have debated how to interpret Thatcher's governments, and many have taken the view that

> some things are crystal clear. First, she had utterly transformed
> Britain's standing and reputation in the world . . . She had also been
> responsible for changing the whole climate of British politics by
> emasculating the political power of the trade unions and returning

Designing Modern Britain

vast sectors of the economy to private ownership . . . From 1982 also, she could boast that the British growth rate was the highest in Europe and that the increase in productivity was the highest too.[11]

Others contrasted the political stability of the 1980s with its economic instability, noting that the worst post-war slump occurred between 1988 and 1992. For those in work, there was a 'rapid and sustained rise in earnings', but for those without there were record levels of unemployment in this period, up to 11.5 per cent of the population in 1993, and an under-class emerged that was economically and educationally underprivileged.[12] Even so, the predominant assessment of the period 1979–92 was decisively negative: 'for a short period between 1985 and 1988, it seemed just possible that Thatcher's governments ha[d] succeeded in reversing Britain's relative decline.'[13] By 1990, certainly by 1992, that claim was no longer credible. Instead, 'her period in office would be remembered as one that began and ended with recessions, the first the deepest, the second the longest in post-war British history.'[14] Even those at the heart of the new enterprise culture faced difficulties:

> Small businesses were devastated by the second recession, and many new homeowners were to have their homes repossessed. Inflation and unemployment once again rose to desperate heights. Compared with foreign competitors, Britain seemed to be doing very badly.[15]

This discussion of the significance of Margaret Thatcher's governments may seem somewhat prolonged, but in examining design in the 1980s and '90s Thatcherite doctrines were pervasive. The comedian Harry Enfield's comedy TV character 'Loadsamoney' was a grotesque parody of the get-rich-quick culture of the 1980s, delivering large sums to those in some jobs (typically, not those in manufacturing and the public sector, but rather those in the service sector and in the property and financial markets) and symbolized by the acquisition of designer 'things'. Individual consumption rather than mass-production dominated the late twentieth century; the emphasis was to get ahead on your own so as to reap personal rewards. But those who were economically disenfranchised lacked the capacity to consume. For them, there were only 'the "slavery" and indignity of poverty, the imposed loss of identity, the almost "no person" status of those not able to make meaningful market choices or even present themselves as potential buyers.[16] Race was another arena in which Conservative policies wreaked havoc, with riots in 1981 in Liverpool, London and several other British cities, as well as numerous pieces of racist legislation to limit immigrant numbers throughout the 1980s. In such a context, heritage reinforced particular identities – predominantly white, male and, following the Falklands War in 1982, 'Imperial British'. Rather surprisingly, since Thatcher's brand of conservatism had ditched a number of traditional 'Conservative' values, the heritage industry 'popularised and commodified the values of a more ancient, patrician and rural Conservatism'.[17] It also repackaged industry and manufacturing, often on the very sites where real industry had been based, but inevitably minus the dirt. In parallel, distinctive local and regional cultures, styles and crafts, took on particular significance within the context of global, standardized products and markets, since 'the struggle for place is at the heart of much of contemporary concern with urban regeneration and the built environment'.[18] Hewison cited several examples, but Beamish, the North of England Open Air Museum in County Durham, stood out, telling

> the story of the people of North East England at two important points of their history – 1825 and 1913. In 1825 the region was rural and thinly populated. The industrial revolution, especially the coming of the railways, accelerated change. By 1913 the region's heavy industries were at their peak.[19]

Beamish Museum included several historical reconstructions of everyday life – working-class homes, a co-operative store, a mine, a farm and a dentist – so as to evoke and represent a sense of 'place'. Its website in 2006 explained:

> Beamish is not a traditional museum. Most of the houses, shops and other buildings have been 'deconstructed' from elsewhere in the region and rebuilt here. A few, the Drift Mine, Home Farm and Pockerley Manor, were here already. All are buildings filled with objects, furniture and machinery – real things from our extensive collections.[20]

Inevitably, the complexities of life in the north-east of England are reduced to a number of selected narratives, but these have been grounded in historical research, archival sources and the recollections of local people.

In the second half of the 1990s, after four election defeats, the political direction of the Labour Party began to move from left to right. To appeal to the new property-owning, service-sector-employed middle class, it undertook several crucial reforms; most controversially, its constitution was rewritten, breaking the historical commitment to 'the common ownership of the means of production, distribution, and exchange' (clause IV).[21] Added to this, its relationships with the trade unions were weakened, and their power to elect MPs was reduced. Not surprisingly, the commitment to enterprise was reiterated, although this was set alongside more traditional Labour commitments to social justice. This rather set the tone of 'New Labour', which, following a landslide election in 1997, gave Tony Blair's Labour Government a huge 178-seat majority. Continuing many of the policies begun by the Conservatives, the new government wanted individual entrepreneurs, but it also aimed for 'a society where we do not simply pursue our own individual aims but where we hold many aims in common and work together to achieve them'.[22] This seemed to indicate a distinct difference from previous Conservative governments, but by 2001, when a second Labour Government was elected with another large majority, it seemed that Labour's greatest achievement was to have 'efficiently husbanded the same trends forward'.[23] Reflecting on this in relation to design in Britain, there was a perception that the strident individualism and obsession with particular forms of the 'English' past typical of the Thatcher years had been replaced by a more diffuse set of cultural reference points, framed by a concern for modernity and technology alongside a sensitivity to diversity. The various millennium projects were symbolic of the former, in particular the much-maligned 'Millennium Dome', but also the London 'Eye' Ferris wheel and the double-helix footbridge over the Thames. Modernity and technology are also in evidence in the 'Blinking Eye' Millennium footbridge over the River Tyne, connecting Gateshead and Newcastle. This evokes a sense of diversity through place by referencing Tyneside's engineering tradition and nineteenth-century technology-led industries such as Armstrong. At the same time, historic buildings, such as the former Baltic Flour Mill and the newly constructed Sage Music Centre on the south bank of the Tyne, have been marshalled in the cause of modernity and culture with a nod to heritage.

Design and Enterprise Cultures

In the 1980s London led the way in harnessing design to enterprise and heritage initiatives, particularly through urban renewal projects, but this strategy was also evident in the regions, where it took on different forms. Influenced by American and European examples, planners tried to revitalize city centres and depressed regions by developing cultural activities in tandem with new housing, shopping and cultural institutions. Instead of the ambitious modernist urban plans of the 1960s and '70s, these initiatives were typically spearheaded by one or two distinct buildings – a shopping centre, public building or a gallery or museum – and design was an integral feature. Lumley suggested that the museum was 'undergoing a "renaissance". In Britain, new museums are being set up at the rate of one a fort-night.'[24] Equally, 'a renewed desire on the part of governments, cities, and private individuals to invest in museums', contributed to 'the museum as architectural innovation and hub of urban redevelopment; [and] the rise of the museum director as star'.[25] These new cultural districts in British cities were not simply places to be visited, or sites solely for the consumption of an array of goods and services, but they were also involved in making and re-making the past. Consequently these were

> a political resource whereby national identities are constructed and forms of power and privilege justified and celebrated; alternatively, or additionally, they have pointed to the resistance and opposition expressed through recourse to other traditions.[26]

Looking back on this phenomenon from a vantage point in the first decade of the twenty-first century, it is useful to consider what forms this re-making of the past took, and the extent to which there has been any evidence of a recourse to different narratives or traditions. A clear trend has been the adaptation of historic premises and/or the redevelopment of derelict industrial or brownfield sites. Design, art and other cultural institutions have been rehoused in defunct industrial buildings or parts of buildings. An early example was the adaptation of disused space in the

Quayside, Newcastle upon Tyne, showing the Gateshead–Newcastle Millennium Bridge designed by Wilkinson, Eyre Architects, the Baltic Centre for Contemporary Art (a former flour mill) designed by Ellis Williams architects and the Sage Music Centre designed by Foster and Partners.

Designing Modern Britain

Victoria and Albert Museum to provide a new gallery for the exhibition of twentieth-century design; the Boilerhouse opened in 1987. More recently, Tate Modern (2000), a vast new art gallery, was developed in Sir Giles Gilbert Scott's Bankside Power Station, built between 1947 and 1963, while Baltic (2002) in Gateshead–Newcastle, designed by Dominic Williams of the architectural practice Ellis Williams, resulted from the conversion of the former 1950s Baltic Flour Mill into an 'art factory'. A process of cultural decentralization took place towards the end of the 1980s, in which major national institutions based in London opened satellite museums in the regions. Tate Liverpool was one example. Opened in 1988, it was located in a Grade 1-listed nineteenth-century warehouse in Albert Dock that was converted by James Stirling. As at Baltic and Tate Modern, this exploited and extended the structural ingenuity of nineteenth-century engineers and builders to develop large spaces for the exhibition of art. Architectural ingenuity and structural innovation also characterized several of the prestigious new art gallery and museum buildings. In 1993 Tate St Ives opened in a new modernist building designed by David Shalev and Eldred Evans, which evoked the forms of the site's earlier gasometer; with stained glass by Patrick Heron, it recalled the collaborative ideals and practices of the early 1930s modernists based in St Ives. Other new art galleries and museums built outside London included the Royal Armouries Museum, which moved part of its collection to Clarence Dock, Leeds, in 1996, while in 2000 Stirling and Wilford designed the Lowry to house the work of the painter L. S. Lowry; built on Salford Quays, it looks across to the new Imperial War Museum North (2002) designed by Daniel Libeskind. Several of the designs – for both conversions and new structures – combined heritage and modernity, mixing the traditional with contemporary forms and materials. Baltic, the Lowry, the Imperial War Museum North, Tate St Ives and Tate Liverpool promoted their modernity as much as their 'heritage'. The Lowry, for example, used glass and metallic surfaces to reflect the surrounding landscape and waterways; the Imperial War Museum North clad in aluminium used new technologies to depict a 'globe shattered by war and conflict'.[27] This building, like Baltic, incorporated spectacle into the design with high glass and open metal viewing platforms, to give dramatic views across the Manchester Ship Canal.[28] At Baltic, only the south and north facades were from the original 1950s building; the interior was an entirely new structure with two new facades providing 3,000 square metres of arts space (four galleries and a flexible performance space), as well as a stunning glass viewing platform overlooking the River Tyne. Typically, these new buildings incorporated artists' studios, cinema/lecture spaces, shops, libraries and archives, and restaurants and coffee bars. Here culture and commerce came together, because shops, restaurants and bars were essential ingredients in their economic survival. Most of these museums and galleries are to be found on derelict industrial sites, including docks (Albert Dock and Clarence Dock), quaysides (River Tyne), canals (Manchester Ship canal), defunct industrial buildings (Baltic, Bankside) and miscellaneous brownfield sites (St Ives).

In part, these celebrate industrial heritage, but fundamentally they belong to a process of urban renewal that drew together the wider economic regeneration strategies of central government, local authorities, the cultural sector and private investment. Such strategies have been criticized for absorbing large quantities of public sector resources that might be better spent on improving the quality of life for local residents.[29] While this may be true for some areas – perhaps the early developmental phase of London Docklands is a case in point – there were important city-centre sites that, dismal and derelict, were in need of significant new landmarks. Gateshead–Newcastle quayside is an example. Over the past seventy years or so, Tyneside has seen a variety of forms of enterprise cultures, ostensibly to arrest economic decline. Some, such as the *North East Coast* exhibition in 1929 and the building of the Team Valley Trading estate at the end of the 1930s, were undertaken during periods of economic decline. There were also a number of initiatives focused on rejuvenation in the post-war period that did stimulate a 'sense of place' and locality after a period of modernist planning in the 1960s, such as the Byker Wall. But achieving an appropriate balance between the local, the national and the international or global has been difficult, and although it is important not to devalue the vitality of local cultures, it is crucial to avoid the overt nostalgia of heritage sites such as the largely fictional 'Catherine Cookson Heritage Trail' in South Tyneside.[30] More successful have been the new heritage initiatives, such as the open-air museum at Beamish in County Durham, and although this too has attracted negative critical attention, it aimed to evoke 'past ways of life of which the visitor is likely to have had either direct or, through parents and grandparents, indirect knowledge and experience'.[31] Equally, it has provided a unique regional resource for the serious study of aspects of everyday life, material culture and labour history. It appears to exist in an idyllic rural landscape, but it is possible, via the museum's archives, photographs and oral-history accounts, to build up a less cosy image of life in the north-east of England. Beamish represents an attempt at the serious reconstruction of the past, in contrast to the reconfiguring of heritage and modernity that took place in the last decade of the twentieth century along the River Tyne in Gateshead–Newcastle, which aimed at the regeneration of the north-east of England's principal city. With the development of significant public building, revitalization of Gateshead–Newcastle has managed to combine, *by design*, a sense of place with an outward-looking and 'energetic cosmopolitanism'.[32] It included the Law Court; new public infrastructure projects, such as the new 'Blinking Eye' Millennium footbridge over the River Tyne by architects Wilkinson Eyre Architects and engineers Gifford Partners; the formation of new cultural institutions such as Baltic and the recently opened (2005) Sage Music Centre designed by Foster and Partners; and commercial initiatives, such as hotels, bars, restaurants and luxurious 'lofts', but also more modest housing schemes. The centrepiece is the innovative and technologically adventurous Millennium Bridge, but, of course, there is a trade-off. There is scant recognition of the labour, hardship, enterprise and community that previously existed along this stretch of river; Baltic, for example, was formerly a hive of activity for the processing of grain. But the Gateshead–Newcastle quayside is now a lively place, albeit dedicated to the consumption of various services and consumer goods, as well as the acquisition of high cultural values (although Sage is a venue for pop in addition to classical and traditional music). Instead of opting for the reconstruction of one particular aspect of its history (and it would be difficult to choose a particular period, since Newcastle's river was busy in the medieval period and again in the nineteenth century, but activity shifted higher up the river banks in the intervening years), the quayside has been transformed into an eclectic mix that not only refers to the strong engineering traditions of Tyneside, but also its formative role in the ongoing modernization of Britain.

Celebrities, Cults and the Design Museum

Inevitably, specific notions of taste and cultural value prefigured the new museum and gallery developments of the last two decades, and nowhere has this been more apparent than in the development of the Design Museum in London. Its origins lay in the Boilerhouse at the Victoria and Albert Museum, but its opening at the end of the 1980s was timely. In providing space for the display and exhibition of twentieth-century artefacts, it responded to a growing interest in contemporary design in which particular 'designs' were heralded as 'classics' and 'cult' objects; 'designers' were courted as stars and celebrities; and high-profile museum directors were cultivated as social and cultural commentators. The Design Museum was housed in a former warehouse building, and from the outset it attracted criticism. What seemed to rankle most was the museum's close relationship with business and commerce, even though historically design had been perceived as a tool to enhance the economic viability of manufacturers and

The Design Museum, London, opened 1989. A former warehouse transformed into an international-modernist style museum.

the trading capacity of Britain.[33] So why did the Design Museum attract such negative press, particularly at its foundation? Without a doubt it was viewed as a product of Thatcherite values, exacerbated by its location in London's docklands. The establishment and activities of agencies such as the London Dockland Development Corporation (LDDC), formed in 1981 by Michael Heseltine, partly in response to the inner-city riots, were viewed with some concern. With little attention paid to the needs of the local residents, there was a perception that regeneration was in the service of the well-off 'yuppy' incomers.[34] Smart, up-market retail outlets, restaurants, wine bars and warehouse living dominated, instead of affordable housing, public transport infrastructure and better schools and hospitals. The activities of the LDDC, however, did change over the 17-year period of its existence (1981–98), and in relation to housing – a controversial subject – gradually the mix became more 'typical' of the rest of London, with a combination of traditional local-authority housing, social housing often in partnership with housing associations, and private development, although the last was by far the largest sector.[35] Not only was the Design Museum associated negatively with expensive warehouse living and up-market retail developments, it was also seen as an expression of the cultural values and aspirations of its founder, the retailing magnate Terence Conran. Its sparse modernist exterior was matched by minimalist interiors, and curatorially it seemed to conform to the 1980s preoccupation with promoting 'great' names (almost always white, male and Western) and their 'classic' designs. Indicative of this hagiography of designers and the cult of the object was

Designing Modern Britain

Ross Lovegrove's 'iconic' PET plastic bottle designed for Welsh mineral water company Ty Nant, 2001. With its ripple effect evoking the fluidity of water, it was voted in *Arena* magazine the '8th coolest item in the world right now' in October 2003.

Deyan Sudjic's book *Cult Objects*, published in 1985.[36] In this an array of things – the Jeep, the Zippo lighter, the Mont Blanc fountain pen, the Sony Walkman – were labelled 'cult' objects. These mass-produced goods were typically 'modernist', but their status could also be subject to fashion (another 1980s cult object, the Filofax, is a case in point). Although they lacked traditional indicators of status (craftsmanship, luxury and uniqueness), Sudjic proposed that they represented the identity and status of their owners through their 'cult' status, and that consumers had to learn the rules of what constituted a 'cult' in order to communicate with these objects. The designer suit, designer chair, designer this or that became part of popular design culture in the 1980s, and the Design Museum appeared to give authority to this trend.

In the 1990s, however, the Design Museum sought realignment in terms of its aims and objectives, in order to place it more firmly within the realm of the academic as well as the professional design community and the marketplace, and although it still had its critics, it became an important resource for those interested in the serious study of design and its histories as opposed to those of fine art. While its current website maintains the hype, claiming it is 'the UK's cultural champion of design' and it was 'recently ranked among the UK's top five "coolest venues" in a public poll', it nevertheless declares itself committed to 'design's impact on sustainability, inclusivity and quality of life'. Reinforcing this was the controversial award of Designer of the Year to Hilary Cottam in 2005 for her work as head of the Design Council's RED team. By bringing 'together the right people, including designers, policy makers, professionals and the users themselves, to take on seemingly intractable problems within public services', this team adopted a collaborative approach to the design of public facilities, such as schools, hospitals and prisons.[37] Unique in engaging with contemporary as well as historical examples of design, the Design Museum remains essentially modernist in approach, emphasizing progress through the use of new materials, formal experimentation and the application of new technologies. Its website still indexes designer stars and celebrities (Alvar Aalto, Saul Bass, Achille Castiglioni), design classics (Concorde, Aston Martin, Jaguar and Penguin Books) and new materials (plywood and aluminium). The discussion of Concorde, the Anglo-French supersonic aircraft (1976–2003) that reduced the transatlantic flight from London to New York to three hours, was indicative of this approach by stressing innovation and technology alongside futuristic formal qualities. Rather less was said of environmental issues or the high monetary cost to consumers; it was instead cited as 'an exemplar of technological excellence'.[38] Equally, the selection of individual designers and important designs was indicative of particular priorities, particularly formal and technological ones; for example, Jonathan Ive's iMac design for Apple in 1998 and James Dyson's DC01 Dual Cyclone Cleaner of 1993 were cited alongside the work of multimedia designers such as Danny Brown, Martin Lambie-Nairn and the collective Tomato (formed in 1991), which worked for companies such as Sony, Channel 4 and BBC2.[39]

In some respects, the Design Museum's preoccupation with the individual is surprising given the sustained critique over the last 20 years of the idea that the author (read here, the designer) determines and fixes the meaning of cultural products. But paradoxically this critical questioning was paralleled by the rise and rise of the designer as 'star'. The imprint of the designer's name has been an increasingly essential element in the promotion of design as high culture and of everyday things as 'classics'. While this has been posi-

Glo-Ball (1999–2005), produced by Flos (new version with white lacquered stand), designed by Jasper Morrison.

tive in challenging the primacy of fine art and recognizing that designs embody and represent important cultural values and meanings, it is also clear that it has functioned to create 'distinctions' within a vast marketplace. In fact, the mythology of individual design genius has become commonplace in our everyday culture, and the signature of certain designers has added value to all types of design. To some extent such mythologizing has been intrinsic to design ever since designers' names began to be used to promote and sell goods more effectively, at least as far back as Josiah Wedgwood in the eighteenth century. Reinforced by the modernist tendency to celebrate innovation and individual creativity, this process was further legitimized after 1945 by awards and titles (for either people or goods) such as the Design Centre Index, Royal Designer for Industry Award, Prince Philip Designers Prize and Designer of the Year Award. It gathered pace towards the end of the twentieth century as certain areas of design – fashion, furniture, graphic and product design – became increasingly dominated by the designer as *auteur* and celebrity, within the context of late capitalist economics. Here the designer added significance to a product that might otherwise be only one among many, enhancing its cultural and monetary value, the standing of the manufacturer and its potential lifespan beyond immediate use, for example, by museum curators

Designing Modern Britain

and collectors. In the 1990s a number of British designers found success working in this mode for both international and British companies, and designers trained in Britain's art schools and universities often worked with more success abroad than at home. Certain categories of design were more likely to be sold on a 'name', and although these were usually for individual consumption, large-scale, often controversial architectural projects also gained legitimacy as examples of architects' oeuvres. Consider the buildings and activities of Norman Foster, Richard Rogers, Daniel Libeskind, Zaha Hadid and Eva Jiricna, whose reputations were consolidated by high-profile projects, often abroad. A number of international companies regularly employed freelance designers. Theo Young, Ron Arad, Zaha Hadid, Nigel Coates and Jasper Morrison were commissioned to design a range of products for Alessi (Italy); Michael Young and Jasper Morrison designed ceramics for Rosenthal AG (Germany); Jasper Morrison worked at Rowenta (France), Ron Arad at Kartell and Artemide (Italy), Theo Williams and Michael Young at Armani (Italy), Tom Dixon at Artek (Finland), Jasper Morrison and Ron Arad at Vitra International AG (Switzerland), Eva Jiricna and Theo Young at Liz Claiborne (USA), and David Mellor and Jasper Morrison at Magis (Italy). In fashion, several British designers headed international couture houses following successful independent collections in Britain: John Galliano at Givenchy (1995) and Dior (1996); Stella McCartney at Chloé (1997) and Gucci (2001); and Alexander McQueen for Romeo Gigli and at Givenchy (1996).

At the turn of the millennium it appeared that the language of design was increasingly global, as well as technologically and stylistically innovative. It often used colour, decoration and pattern, but could also be formally

Hotel Josef, Prague, designed by Eva Jiricna, 2002.

Bookworm bookshelf designed by Ron Arad for Kartell, 1996.

minimal with popular cultural references – children's toys and sweets in the case of the iMac. Compared with the grey uniformity of office computers, the curvilinear iMac looked playful and easy to use, with its translucent body in a choice of five 'fruity' colours – Blueberry, Grape, Tangerine, Lime and Strawberry. In contrast, the classically white Apple iPod, (again by Ive from 2001) was a small, portable unit (just over four inches tall and weighing less than seven ounces) that stored and played

Tefal Aquaspeed steam iron designed by Dick Powell of Seymour Powell for Calor, 2004.

James Dyson, DC03 dual-cyclonic vacuum cleaner, 1998. Dyson's revolutionary bagless vacuum cleaners were launched in 1993.

1,000 top-quality music files. Dependent on new technologies, both in terms of production and use, it was made from sealed twin-shot plastic combined with an innovative one-point thumbwheel controller. It rapidly became ubiquitous and was followed by the iPod Mini and the iPod Nano (in a range of colours and personally engraved), their usability enhanced by technological developments, such as iTunes for Windows in 2003.

New technologies have enabled designers and manufacturers to be both imaginative and questioning, for example, the Air Chair designed by Jasper Morrison for Magis in 2000 and the RCP2 Children's Chair by Jane Atfield from 1993. Morrison's chair used gas-injected Polypropylene strengthened with fibreglass to produce a relatively cheap (£60) and durable stacking chair in a variety of colours that could also be used outdoors. Experimenting with technologies – both old and new – contributed to the drive for sustainable design, as seen in Jane Atfield's company Made of Waste, which developed a method of making sheets of pressed plastic from recycled materials such as detergent and shampoo bottles and yoghurt cartons. In the exhibition *Reclaimed Recycling in Contemporary British Crafts and Design*, organized by the British Council in 1999, she wrote:

> I am interested in the idea of reinterpreting disused objects and creating new functions for them. This creates a narrative around these materials and objects. I transform various discarded consumer objects – from plastic bottles to vending cups and yoghurt pots – into construction sheets which can then be used to make new objects such as chairs. This manufacturing process relies on collaboration with industry to develop and research new environmentally friendly materials with a

Designing Modern Britain

The iMac computer designed by Jonathan Ive for Apple, 1998.

The iPod designed by Jonathan Ive for Apple, 2001.

RCP2 Children's Chair by Jane Atfield, 1993. Made from recycled plastics.

distinct aesthetic. Simplicity, functionality and addressing environmental problems are issues that are important to me.[40]

Whereas at first glance the 'new' international style appeared to erode national, regional and political identities so as to exploit commercial opportunities and to address the needs of international elites, designers such as Atfield used technology in creative and politically challenging ways. Others, however, remained engaged with national identities in ever more orthodox ways – evoking nostalgia for an ideal England.

Fashioning 'Englishness': Nostalgia and Identity

Several writers and critics in the 1980s and '90s observed the obsession with a very specific past and the role of 'design' in this: 'It's a place viewed through a cottage window draped in Laura Ashley or William Morris curtains, back-lighted courtesy of Habitat'.[41] Intrinsic to these definitions was a 'story' or a number of 'stories' about 'Englishness', in which identity was shaped by the countryside, by selected historical styles and by a sense of continuity energized by a progressive and exploratory turn of mind best characterized by inventors such as Thomas Telford and designers such as William Morris. In contrast, characterizing 'Britishness' was fraught with difficulties, since Scotland, Wales and Northern Ireland and the English regions sought separate representation. Summing up Britain as 'imperial' resonated in some quarters after the Falklands War, but by the 1990s this was subverted by the diversity of cultural and social identities in Britain. 'Englishness' perhaps evoked a less troubled past, 'retrospectively in soft focus'.[42] Discussing the vogue for heritage in a number of aspects of design – products, advertising and packaging – Jonathan Woodham wrote:

Habitat Chesterfield and ZoZo sofa, *Habitat Catalogue*, 1976.

> There has been a long-standing belief that the quintessential England lay in the countryside rather than the city. The traditional 'unpolluted' values of a rural heritage and a spirit of nostalgia located in a seemingly familiar yet historically unspecified past.[43]

Among the numerous manufacturers and designers to draw on the 'rural heritage' were Johnson Brothers

Eternal Bow, Johnson
Brothers Ltd, designed by
Serena Mascheroni, 1980s.

and Staffordshire Potteries Ltd.[44] Both companies exemplified the ways in
which ceramics manufacturers used design in company strategies.
Following the relaxation of government controls after the Second World
War, they became mass-producers of ceramic tablewares, and from the
1960s began to employ freelance designers –Mary Quant, for example,
worked for Staffordshire Potteries – to raise the standing of their products.
By the 1980s, in a highly competitive economic context, they produced
tablewares that responded to changing consumer needs. Staffordshire
Potteries introduced oven-to-table 'rustic' stonewares in 1982 that were also
suitable for use in the microwave oven. Taken over by Coloroll in the mid-
1980s, the firm was one of the largest manufacturers of mugs in the world,
producing more than 750,000 per week, many of them with commissioned
designs for promotional, celebratory or company purposes. New shapes
and patterns were frequently based on historical precedents; for example,
Johnson Brothers' extremely popular 'Eternal Bow' designed by Serena
Mascheroni incorporated Regency styles and colour schemes.[45] For many
ceramics manufacturers in Stoke-on-Trent at the end of the twentieth
century, economic viability was precarious in the face of new foreign producers, and Coloroll went into receivership in 1999.

The eighteenth century in particular remained an accepted benchmark
of 'good taste' and it was deployed with some regularity by manufacturers,
but the late nineteenth-century, Tudor and various vernacular styles and
materials were also popular for housing, fashion, textiles, ceramics, advertising, packaging and furniture. Neo-vernacular influences predominated
in housing, both private and public, from the 1970s onwards. An important

'Lecontree' private middle-class housing by the builders Bussey and Armstrong, Darlington, 2006.

source for these ideas was the *Design Guide for Residential Areas* produced by Essex County Council in 1973. Initially aimed at housing developers who showed little sensitivity to local vernacular traditions, it encouraged builders to consider housing in relation to existing older buildings, particularly those from before the First World War, and to be responsive to local traditions in materials, forms and detailing. The *Design Guide* drew on prototypes from the USA, which had emphasized low-density, small-scale developments of low-rise housing with good landscaping, and became both a rulebook for acceptable taste and 'a set of rigid design rules'.[46] The timing of this was very apt, as local authorities, planners and builders were turning away from modernist high-rise building and beginning to search for something more modest and appropriate, and perhaps 'English'. The plethora of developments of individual, semi-detached or small terraces set in culs-de-sac with integral planting that developed across the country in response to these guidelines were dismissed by many architectural critics. Sutherland Lyall, editor of the highly influential *Building Design*, was scathing: 'Dull, hack with a few tricks attached, quasi-archaeological, highly wrought or semi-Disneyland, Neo-vernacular is the reigning style in mass housing.'[47] By the 1990s, however, a specific set of historical/vernacular design elements was increasingly common, and these referenced particular styles as well as vernacular modes – especially influential was late Victorian. An example is the Darlington firm, Bussey & Armstrong, a recipient of the Daily Mail Green Leaf Award in 2001 for environmental excellence and a Royal Institute of British Architects commendation in the same year for 'cre-

Designing Modern Britain

ating a feeling of identity and sense of place' through design. Established in 1902 to build houses for the town's railway and engineering workers, the firm was also inspired by the Garden City movement. Developing a garden suburb to house workers at the new chemical plant on the north-west edge of the town in the late 1920s, its founder, A. B. Armstrong, was committed to the provision of affordable homes at a cost of £400. With a living room, kitchen, scullery and three bedrooms, the houses had modern conveniences, including electricity. The company website today emphasizes the firm's heritage and its long-standing involvement in the provision of housing in Darlington, stressing that at the height of the Depression they used local materials.[48] From the late 1980s the company began to reposition itself in the market, building more expensive houses in the west end of the town and using 'authentic' materials such as 'real Welsh slate, stock bricks and leaded glazing' combined with 'vernacular', 'historical' and 'English' architectural features.[49] The RIBA-commended housing at 'Woodland Park' was set in the grounds of a demolished Victorian villa, and 'Leconfield', within walking distance of the town centre, provided 'high-quality, high-specification town houses and apartments'.[50] These latter houses are an eclectic mix of Arts and Crafts and Queen Anne Style architecture and combine roughcasting, red brick with barge boarding, over-hanging eaves, bay and oriel windows, and red pantile roofs. Emulating the practices of Arts and Crafts architects and designers, the company employed an artist-blacksmith to make finials and railings.[51] The designs typify the use of vernacular and historical styles in housing in the last 20 years, and in this particular area of Darlington they are typically small-scale infill developments that butt up against original Georgian, Victorian, Edwardian and interwar housing to evoke a sense of continuity, tradition and 'Englishness' in a town that has largely escaped the worst effects of modernist 'grand planning'.

In his sketchy celebration of the 1980s Peter York pointed out that the property boom made people aware that there were vast swathes of existing housing stock out there, and that crucially it could be restored and modernized.[52] Additionally, for those who wanted to buy new housing, there was Docklands,

> the apotheosis of London's redeemability. While it was one thing to reclaim North Clapham, with its large, sensible houses and its tolerable infrastructure, it was an imaginative leap of a different order to reclaim a load of huge, rotting tea and coffee warehouses in a part of London which was not only virtually off the map . . . but quite impossible to get to.[53]

Ironically, although the new apartments built in the old Docklands warehouses were externally traditional in style and materials, internally they were strikingly modern. The metal-framed structures permitted huge open spaces and vast expanses of glass that were combined with state-of-the-art kitchens and bathrooms to produce a curious mix of heritage and contemporary styles. Interior design and furniture companies became adept at producing a variety of historic styles, partly in response to the demands of those restoring the numerous nineteenth-century terraces, and semi-detached villas. Kitchen manufacturers excelled in producing a range of looks for different markets: 'English' country scrubbed, traditional craft bespoke and hi-tech. In terms of furnishings, straightforward reproductions were still available, but more popular was an eclecticism borne of 'collecting' junk and antiques, or 'repro' versions of originals reworked to give a 'modern' feel. Habitat continued in this vein, producing, for example, Chesterfield sofas based on Victorian designs, but combined with modern fabrics. These styles and tastes perpetuated those of former ruling elites, although the vogue for scrubbed, 'distressed' country pine connoting the rural poor proved remarkably resilient. At the same time, organizations such as the National Trust began to sell its own range of traditional 'English' textiles, ceramics and other household goods. Its presentation of historic houses amounted to a celebration of past values in the present, in particular 'a respect for privacy and private ownership, and a disinclination to question the privileges of class'.[54] To a considerable degree, the National Trust's approach was ahistorical; not only did it aim to present the house as though the current owners had 'just popped out', it often borrowed items to construct specific narratives that depicted an unchanging and unproblematic patrician 'Englishness'.

In an attempt to identify something of the dynamism and changing character of 'Englishness', the architectural historian Nikolaus Pevsner – perhaps with his own origins in mind – pointed out that 'England has indeed profited just as much as from the un-Englishness of the immigrants as they have profited from the Englishing they underwent'.[55] Discussing English art, Pevsner's central thesis was that Englishness was a struggle between two opposing sets of qualities or characteristics: 'on the one hand there are moderation, reasonableness, rationalism, observation and conservatism, on the other there are imagination, fantasy, irrationalism'.[56] He concluded that England lacked genius in the visual arts because of the ascendancy of what he termed the 'English' qualities of practical sense, reason and tolerance over those of fanaticism and intensity, which he attributed to great art. It has become de rigueur to discuss the work of Vivienne Westwood in terms of 'Englishness', but reading Pevsner this does indeed seem apt. Westwood's approach to fashion involved imagination, fantasy and irrationalism, and irrespective of claims made for her genius, she responded and contributed to the process of questioning about identities in the late 1980s and '90s. Her self-presentation was deliberately intellectualized and critically engaged, although she also emphasized the 'craft' of making clothes. In discussing Westwood's work, McRobbie observed that she embraced a fine-art mode in her approach and 'presents herself with an air of practised eccentricity, which is of course a recognisable and accepted way of being an artist'.[57] Her early work was formed within the context of Punk, which, in addition to scrutinizing gendered identities, represented a clear identification with black British and West Indian culture and style (Pevsner's un-Englishness borne of immigration), but also alternative traditions of dress and subversive readings of class and history. Collections such as Pirate were highly feminized and decorative, implying contested sexualities, and its bold colour, brightly-patterned prints and layered clothes referenced non-white European cultures, whether black British or African. This had an added currency during the 1980s and early 1990s, when designer labels, power dressing and the resurgence of interest in royalty (after Charles and Diana's wedding in 1981) were to the fore. In this context, Westwood's collections were in some ways an assault on standards of taste at a time of Conservative politics and conservative values, when to dress appropriately 'became a vital sign of financial success', but 'Westwood, with tongue firmly in cheek, went a dress code further though, and dressed as the Queen'.[58] Various writers have commented that Westwood enjoyed exposing the iniquities of the 'English' class system, but paradoxically she yearned for the refinement and luxury of aristocratic dress from the eighteenth century. 'Englishness' in Westwood's work was defined in class terms – the country house and its cultures.[59] At a time when national and regional identities were vigorously contested and asserted, Westwood conflated 'English' and 'Scottish' in several collections by deploying kilts, fair-isle, tartan, and tweeds with, at best, a superficial engagement with wider political meanings.

In reviewing the obsession with 'Englishness' and its collapsing into 'heritage' in the 1980s and '90s, it is worth noting that 'Britishness' enjoyed less prominence. This may have been caused by the political context at the end of the twentieth century, as demands for devolution in Scotland and Wales gathered pace, but 'Britishness' also required an uncomfortable confrontation with the unpredictability of geographical extremities – both within and outside Britain, in particular, the legacies of 'Empire' and subsequent immigration, but also the complexities of Irish, Scottish and Welsh nationalism. Equally beyond the bounds of this particular cosy and rural manifestation of 'Englishness' was the 'metropolitan' south-east of England. Pevsner's ideas are worthy of further attention in respect of design over the last twenty years, not least because his conception of 'Englishness' was grounded in the 'visual' realm, and concerned less with its histories than its geographies. It is perhaps ironic that it is in fashion design that some of the more interesting critical engagements with 'Englishness' took place in the 1980s and '90s, even though this is intimately linked to global markets. What is of particular interest, then, is why and how certain notions of 'Englishness' had any credibility in such a marketplace, and what was the role of design in this process. Recently the dress historian Aileen Ribeiro considered Pevsner's views, arguing that it is almost impossible to pin down what makes a design English.[60] Fairly confident that 'dress, if it means any-thing at all, concerns itself with social norms', she observed that 'although they may be modified by individuals, reflect the customs and aesthetics of any given age . . . our usage of clothing is rooted in the complexities of English cultural history'.[61] Without a doubt, these complexities accelerated as the millennium approached and designers, manufacturers and con-sumers became more articulate in exploring, expanding and questioning the meanings of design. Writing about tradition and style in contemporary British, rather than English fashion, Catherine McDermott pointed to the interest in the countryside and the use of practical durable garments as indicative of 'British' style. But rather in the manner of Pevsner, she argued that there were two opposing tendencies in 'British' identity, 'a taste for practical, simple garments was always partnered with a rather different consciousness, the idea that clothes could express independence of mind, radical social values, even revolution'.[62] Indeed, one might argue that fash-ion has contributed to the formation of numerous identities in the last 20 years, and to paraphrase Ribeiro, these *represent* (my emphasis) the customs and aesthetics of late twentieth-century Britain. But although national iden-tities have been central, a range of other identities has been formed, reproduced and contested.

New Subjects, New Identities

A defining feature of the 1990s has been the sustained critical interrogation of the position of the white male subject as the keystone of human experi-ence. The reasons for this are diverse, and not easy to sum up, but without doubt the impact of 'identity politics' has been crucial. Questions of race, culture, gender and generation have contributed to existing debates about class, but they have also led to new enquiries and alternative strategies. There has also been a sustained process of academic debate provoked by the

Promotional card by
Matrix, showing the
Jagonari Centre and
Jumoke Nursery,
late 1980s.

theoretical ideas emanating from feminism, postmodernism and post-structuralism. This process of critical questioning has responded and contributed to the immense social, economic and political changes that have taken place in post-war Britain. Equally important has been the changing political landscape of Britain and the attitudes of both right and left. Writing at the end of the 1980s, Chapman and Rutherford identified, in particular, the moral indignation of the right; they drew attention to the promotion of Victorian values by Margaret Thatcher; and they highlighted the stigmatization of 1960s permissiveness by Conservative Party politicians such as Norman Tebbit. But, significantly, they also pointed to the inability of the left to deal with the new political matrix. As David Harvey argued, 'we can no longer conceive of the individual to be alienated in the classical Marxist sense, because to be alienated presupposes a coherent rather than a fragmented sense of self from which to be alienated.'[63] He proposed that the fragmented postmodern subject had displaced the alienated subject of Marxism:

> If, as Marx insisted, it takes the alienated individual to pursue the Enlightenment project with a tenacity and coherence sufficient to bring us to some better future, then loss of the alienated subject would seem to preclude the conscious construction of alternative social futures.[64]

Yet one can argue that in the 1980s and '90s new subject identities contributed to the construction of a range of 'alternative social futures', although these were neither modernist nor Marxist. An excellent example was the London-based feminist architectural co-operative, Matrix, established in the mid-1980s with the aim of involving the user of buildings in the design process:

> This approach which favours co-operation over inflexible professional authority and collaboration over inflated individuality makes architecture accessible to clients who have limited knowledge about the design process and who are not normally represented in the building process.[65]

In their *Making Space: Women and the Man-made Environment* of 1984, Matrix explained their approach: 'We are women who share a concern about the way buildings and cities work for women.'[66] The origins of the group were in radical practice and politics developed in the 1970s under the aegis of the New Architecture Movement, a group of socialist architects, students, teachers and builders. Matrix's practice involved exhibitions that addressed architectural and design issues relating to women's lives, such as housing. They also contributed to conferences, wrote articles and designed buildings in order 'to develop a feminist approach to design'.[67] As architects, they worked very closely with their female clients, helping them to find and assess potential buildings and sites; they gave them assistance in obtaining funding for construction; and in order to draw their clients – many of whom had no experience of working with an architect – into the process, they made models and ran workshops to help women's groups understand and take part in design. They believed that, since women were the primary carers in society, 'unless a conscious effort is made to the contrary, the environment will be designed from a point of view that ignores not just women, but also those, such as children or the disabled that are in their care.'[68] As part of this, Matrix were directly involved in campaigns around women's safety; in *Making Space* they highlighted the problems for women posed by urban and new-town designs planned primarily with the car-user in mind. They worked closely with the public sector in the 1980s, but by the 1990s such work was harder to secure, with changes in funding and the sector's diminution in size. Design projects included the Jagonari Women's Educational Resource Centre in Tower Hamlets and the Jumoke Family Nursery and Training Centre in Southwark, both in London.[69] Both projects were closely linked to the local communities, and in the case of the Jagonari Centre, Matrix worked with different groups of British Asian women, for example, helping Somali and Bangladeshi women living in temporary accommodation to develop language and practical skills. It also provided a playgroup, and advice and support on education, training and employment. Designing the Jagonari Centre, Matrix involved the users in the design process, and they attempted to introduce elements into the design that were relevant to the wider cultural context of the women's lives. For example, a decorative external window grille (essential because of racist vandalism in the area) was designed with motifs drawn from traditional Asian architecture, and although the interior was essentially modernist with flexible spaces to facilitate multi-use, there were nevertheless visual references to the women's different cultural identities.

A concern for 'difference' was clearly discernible in fashion retailing in the 1980s and '90s, as new markets were identified, if not created. A company leader here was Next, which recognized new market segments and then set out to design for these. Established by the designer George Davies in 1982, the name Next implied change and difference, as did the various elements of the concept: Next for Men (1984), Next Accessories and Next Interiors (1985), Next Boys and Girls, Next Too, Next Collection (1987) and Next Directory (1988). Initially, it offered smart, fashionable clothes for women over 25, who to some extent had been neglected by other retailers, since the focus was usually the youth market. This group of women with a high disposable income and a continuing interest in fashion was also increasingly discerning about fashion.[70] Suits, separates and accessories were produced in part in recognition of the purchasing power of these ageing baby boomers, who, benefiting from increasing economic autonomy, wanted smart work clothes at affordable prices. Davies, however, also realized that there was much more that could be achieved with 'segmented markets' and the articulation, through design, of particular lifestyles. Next stores were smart and stylish with modern design features – chromium steel staircases, simple timber floors and wall units, subdued lighting – often with coffee shops, offering '"quality" merchandise, in quality surroundings, yet aimed at the mass consumer'.[71] Within this context it nevertheless managed to convey a feeling of exclusivity, since 'style stories' composed of small groups of coordinated items were displayed together. The stores were never cluttered with huge amounts of stock; instead new Electronic Point of Sales (EPOS) systems and 'just-in-time' distribution techniques 'meant that stock could be stored in low-rental, out-of-town warehouses, and supplied instantly when needed'.[72] This also enabled large, unpopular production runs to be avoided. Described as post-Fordist, this strategy involved the promotion of a distinctive up-market lifestyle, and it relied upon new computer technologies.[73] These new retailing strategies were accompanied by equally distinctive graphic design in the form of packaging, carrier bags, promotional leaflets and advertising, and typeface. The clothes provided for a range of 'identities': young and up-to-date, classic style, casual and easy-to-wear, and business smart. The first Next Directory, in 1988, had sections for 'Classic Women' that included, for example, a standard beige raincoat, navy-blue double-breasted pinstripe suit and classic pleated skirt. Next Gentleman, which claimed to have 'rightly established itself for the supremacy of its Tailoring', included simple striped shirts, jacquard tie and a flannel-mix suit.[74] The Directory included fabric swatches of suit materials to give an up-market feel to a mail-order business. It had additional sections for all the segments of the Next market: 'Contemporary women', 'Modern Man', 'Next Boys and Girls' and 'Interiors'.

The launch and success of Next Man responded to the changing roles and perceptions of men, particularly towards the end of the 1980s and in the 1990s as a 'new man' apparently emerged. Partly a product of advertising and marketing, it was nevertheless indicative of men's shifting position in society, as

> the changing nature of work, and the disruption of work culture
> with the decline of manufacturing industry, the introduction of new
> technologies and the subsequent deskilling of traditional male jobs . . .
> undermined traditional working-class masculinities.[75]

The roles of middle-class men changed along with those of women, as feminism influenced gender relations. As women tried to combine challenging

Designing Modern Britain

jobs with having children, men were expected to take a fuller role within the family, both practically and emotionally. Frank Mort argued that the scrutiny of masculinity that occurred in the 1980s partly prompted the launch of new advertising agencies that helped to develop a new approach to young men.[76] There was also a sustained questioning of men's relationships with each other, particularly in relation to gay identities, and 'the current eroticisation of men's bodies, the shifting of gay erotic images into mainstream culture, represents a blurring of sexual differences and a loosening of masculine rigidity'.[77] Powerful advertising images in new magazines such as *The Face* (launched 1980) and *Arena* (launched 1986) contributed to 'an image of sexual diversity, one that traditional masculinities have vehemently denied'.[78] These magazines represented masculinity not as singular but plural, and in so doing they recognized and encouraged change.

Photography and visual style were crucial elements of these new magazines, and, as with Next, helped to define the alternative identities, specific images and design artefacts they depicted. A number of technology-led designs gained iconic status within this style culture (the Sony Walkman first introduced in 1979, for example). Directed at global niche-taste cultures across the globe, rather than local markets,[79] this clearly articulated Saatchi and Saatchi's proposition that 'there are more social differences between midtown Manhattan and the Bronx than between Manhattan and the 7th Arrondisement of Paris'.[80] Equally, people like Jonathan Ive, designer of the iMac who was awarded a CBE (Commander of the British Empire) in 2006,[81] became actors in a 'star' system reaffirmed by cultural institutions such as the Design Museum and within style magazines.[82] Style, an essential feature of the magazines, was dependent upon the things that one consumed: fashion, books, films and food, and these replaced or coexisted with music and sport in the cultural landscape of some men. Inevitably, this concern for style and visual appearances was also evident in the magazines' art direction, since graphic designers such as Neville Brody rejected the functional, problem-solving approach to graphic design taught in art schools in the 1970s and took a more 'expressive, fluid and painterly approach' to typefaces, combining them with powerful, emotive photographic imagery to evoke a mood or attitude.[83] These provided a range of

Next 'Classic' women's wear in first *Next Directory*, 1988.

Next 'Classic' menswear in first *Next Directory*, 1988.

highly desirable masculine images, and towards the end of the twentieth-century a new magazine culture was helping to facilitate the trend by which men were buying consumer goods traditionally associated with women, such as clothes and cosmetics.

If Next Man was the personification of new masculinities in the 1980s and '90s, then Next Interiors was there to provide appropriate goods for new lifestyle aspirations. As the first Next Directory proclaimed in 1988, 'the Next 48 hours could change your lifestyle'.[84] Furniture, kitchen utensils, glass and ceramic tableware, stainless steel, electric coffee grinders, expresso coffee makers, textiles, lighting and decorative ceramics were available in high-street stores or delivered directly to your home. Next Interiors was 'more than a single co-ordinated group, it is a collection of good individual designs which can live together to fit easily into your existing home'.[85] A good example from the first Directory was Richard Sapper's Whistling Kettle designed in 1984 for the Italian company, Alessi, a product described revealingly as 'more looked at and photographed than actually used'.[86] The components that made up the Next Interiors' 'look' were diverse and eclectic: modern and retro-style ceramics (inspired by Russell Wright's 'American Modern' range from the 1930s), craft-influenced vases and bowls (with floral hand-painting and oriental crackle glazes), classic sofas ('Wordsworth' and 'Cheltenham' ranges, for example) and contemporary wallpapers and textiles. Dining furniture was modern and minimalist, whether in black ash, smoked glass or black metal. On offer – on the high street and via the Directory – was a range of items of contemporary 'good' design that could be combined to create an appropriate 'lifestyle'. Next Interiors' methods owed a great deal to Habitat (significantly, Terence Conran was a non-executive director of Next), particularly in composing small room settings and in sourcing a diverse mix of products.

Indicative of the move away from the high street was the development of retail parks and out-of-town shopping centres, which brought distinctive types of retail development. Huge warehouse-style stores sprang up in the former, and a pioneer of this type was the Swedish company IKEA. More than any

First issue of *Arena*, 1986.

Designing Modern Britain

subsequent retailer, IKEA developed the 'lifestyle' selling approach of Habitat and Next, but it did so from vast stores positioned alongside motorways, within striking distance of large urban areas. IKEA, begun in Sweden in the early 1950s, had stores in continental Europe by the early 1970s; and in 1987 it opened its first British store in Warrington. It was claimed upon its opening in Britain that due to 'its considerable buying power [it] can source from all over the world to provide superior quality products, considerably undercutting on price anything currently on offer in the UK'.[87] Like Habitat, IKEA sold 'lifestyles' and 'looks', but low prices enabled fairly rapid change in response to fashion. But equally, the company claimed that such change reflected broader social shifts within the family structure; as the introduction to the 2002 catalogue put it:

STAINLESS
PRECISION

ALESSI KETTLE
Stainless steel body
with black handle,
brass whistle and pitch pipes,
and copper bottom. A design
classic by Richard Sapper.
Height 19cm

> How do you organise family life,
> when your family no longer neatly
> conforms to the statistical 2+2.5?
> The dynamics change. You have more
> children, maybe a few stepchildren
> into the bargain. They get older,
> their friends sometimes stay over.
> They grow up, their need for privacy
> (and yours) increases. Soon they're
> half in and half out of the nest.
> Accommodate the ebb and flow
> of family life, with flexible sleeping
> arrangements, personal sanctity
> zones and happy family areas.[88]

classic
simplicity

Strongly influenced by 'Swedish Modern', IKEA also adopted other design themes, for example, rural/vernacular/ethnic (rugs) and high-tech (kitchens, office and computer stations). Concerned to stimulate sales in the mid-1990s, the company ran an advertising campaign attacking 'English' taste characterized by 'chintz'.[89] The 'Chuck out your chintz' advertising campaign

IKEA shop interior, Ashton-under-Lyne, Greater Manchester, 2006.

introduced in 1996 aimed to persuade the British consumer to reject quin-
tessentially 'English' interior design features so as to be modern. This direct
assault on a specific symbol of national identity was partly indicative of
the ambiguity of national/international products in a global marketplace.
In design terms, 'Swedishness' meant a particular type of modern design
that originated in the early twentieth century and was synonymous with
a particular formulation of 'good taste', that became increasingly 'international'.[90] Understated, simple, using natural materials, but drawing on
new technologies and materials, 'Swedishness' represented an acceptable
modernism that had already contributed to the 'hybridity' of post-war
British design.

Meanwhile, the face of the traditional high street changed dramatically in the 1980s and '90s with the rise of out-of-town shopping centres and
retailing parks. Discussing the city in the 1980s, Bianchini and Schwengel
argued: 'the extraordinary proliferation of out-of-town superstores, shopping centre, multiplex cinemas and other leisure facilities significantly
reduced the need to use the facilities of the city centre.'[91] As an outcome, the
city centre became economically less viable: large stores moved to the low-
taxed out-of-town complexes and small shops went out of business because
consumers preferred to park for nothing at the new shopping malls. The
high street, littered with charity shops and 'pound shops', is making a
comeback, but the implications of late twentieth-century planning mis-
takes will take some time to reverse. Discussing the debate between T. Dan
Smith, the former Leader of Newcastle upon Tyne, and John Hall, the developer of the Metro Centre in Gateshead, Gardner and Sheppard considered

Designing Modern Britain

the extent to which the service sector had provided jobs for Tyneside's unemployed during a period of accelerated de-industrialization. Writing in 1989, at the nadir of Thatcher's economic management, it was clear that

> the decline of Tyneside's manufacturing industries and its burgeoning service sector merely mirrors a national trend. What everyone wants to know is whether the services can mop up all the workers currently being shaken out of manufacturing industry. In other words, can we, as a nation, shop our way out of mass unemployment?[92]

Tyneside provides a good case study for a discussion about the changing patterns and significance of retailing in the late twentieth-century economy. A hundred years earlier, Tyneside's manufacturing sector had employed 45 per cent of the region's workforce; by 1961 it employed 28.89 per cent, and by 1991 12.86 per cent. The service sector (including the public sector) accounted for 13.94 per cent in 1961, and 36.9 per cent in 1991.[93] Tyneside had a strong regional identity with shopping a prime attraction, due to Newcastle's Eldon Square shopping complex (opened 1976) and other well-established shopping in the city's Northumberland Street (in the 1980s its retail rents were the highest outside of the West End of London), but it seemed surprising that a city which in the 1980s had the highest level of unemployment in the country and the lowest household incomes could sustain this. In many ways Tyneside confounded the stereotype of the depressed North in the 1980s and '90s, but the disproportionate amount spent on consumer goods in the region was partly caused by lower rents and low car ownership, the availability of credit and a propensity to spend rather than save.[94] Why, then, did this new form of retailing succeed on Tyneside in the mid-1980s?

The Metro Centre, built on an old industrial area on the south bank of the River Tyne to the west of Newcastle city centre, opened in 1986. It was the biggest shopping development in Europe at the time of its conception and it drew on American and European models, providing extensive car parking, efficient public transport, a range of high-street stores and leisure facilities. The developer, a former miner, John Hall, put great emphasis on his regional and working-class roots. It had, apparently, only one big advantage – its location: 'situated at the heart of Tyneside, the potential catchment was 1.5 million people within a 30 minute drive time, and 3 million within one hour.'[95] In addition, Hall capitalized on the Conservative Government's urban regeneration policies, which led to the establishment of enterprise zones. This stretch of Tyneside, designated in 1981 one of the first enterprise zones in the UK, brought with it government grants, few planning restrictions and a rate-free status for ten years. It also benefited from Tyneside's new A1 Western by-pass, which effectively delivered shoppers to the shops. The exterior was designed as a huge brick and steel-framed structure with little to distinguish it architecturally except functionalism. The interior design combined heritage with various 'themes'. Typically postmodern, it included: the Studio, the Forum, the Village and the Garden Court, providing a 'film-inspired dining area packed with the very best in shops and restaurants'; a Roman 'forum' with marble floors, classical columns and pediments offering designer fashions; and a taste of 'old England' with Victorian-style detailing and a variety of traditional shops.[96] Cultural eclecticism overlaid with heritage and tradition have become common features of shopping centre design, but as at the Metro Centre, these clearly aimed to differentiate and articulate different markets and to provide variety rather than homogeneity. Promotional literature claimed to stimulate family shopping, as people were encouraged to consider shopping a leisure activity. In addition, advertising and promotional literature was published in Norwegian and Dutch to appeal to the North Sea ferry market. From the outset, the Metro Centre was 'managed' very closely. High levels of internal security controlled litter, graffiti, crime and vandalism; and although a combination of both private and public space, the latter was both controlled and confined to 'certain locations, certain hours, and certain categories of "acceptable" activities'.[97] In 2004 a massive expansion of the Metro Centre was opened, making it once again the largest in Europe – in part in response to competition from other local retail initiatives on Tyneside and across the region. Following the unveiling of *The Angel of the North* by Anthony Gormley in 1996 and its perceived

Metro Shopping Centre,
Gateshead, 1986.

Designing Modern Britain

success in the cultural and economic renewal of Tyneside, the new extension also included public art works. The initial development of the Metro Centre appeared, at the time, to represent a challenge to the economic viability of Newcastle city centre, but this had not occurred by the end of the 1990s. Partly this was ameliorated by the rehabilitation of the Quayside area, which helped to reorient the city. Tyneside may provide a paradigm for the shift from production to consumption that characterized the post-war period, but one outcome has been a fairly radical change in its self-image and identity.[98] Perhaps unsurprisingly, this is epitomized by the design of the new Millennium footbridge crossing the Tyne. Here is a design that builds on the technological innovations and modernizing impulses of nineteenth- and twentieth-century Tyneside, thus evoking a strong sense of place, but within a language of modernity, not heritage.

At the end of the twentieth century, design was diverse and multifaceted. Formed within a framework of de-industrialization, enterprise and preoccupations with 'Englishness' rather than 'Britishness', it is worth recalling Pevsner's observation about the role of the 'unEnglish' in stimulating the cultural dynamism required to dislodge an obsessive recourse to idealized and highly partial readings of the past. Perhaps it is overly optimistic to celebrate Tyneside's Millennium footbridge as both modern and sensitive to changing identities? But while noting the power of design to reinforce social inequalities and conservative cultural values (engendering, among other things, nostalgia for a fictional 'England'), we have also seen, in the myriad of design in Britain at the end of the twentieth century, that there has been scope for critical questioning.

References

Introduction

1 M. Berman, 'The Experience of Modernity', in *Design after Modernism: Beyond the Object*, ed. John Thackera (London, 1988), p. 36.
2 David Gilbert, David Matless and Brian Short, *Geographies of British Modernity: Space and Society in the Twentieth Century* (Oxford, 2003), p. 4.
3 For discussions of national identity, see Gilbert, Matless and Short, *Geographies of British Modernity*; Dave Russell, *Looking North: Northern England and the National Imagination* (Manchester, 2004); Nikolaus Pevsner, *The Englishness of English Art* (London, 1955); David Peters Corbett, Ysanne Holt and Fiona Russell, *The Geographies of Englishness: Landscape and the National Past, 1880–1940* (New Haven, CT, 2002); Robert Colls and Philip Dodd, *Englishness: Politics and Culture, 1880–1920* (London, 1987); Benedict Anderson, *Imagined Communities: Reflections on the Origin and Spread of Nationalism* (London, 1983); Robert Colls and Bill Lancaster, *Geordies: Roots of Regionalism* (Edinburgh, 1992); Robert Colls and Bill Lancaster, *Newcastle upon Tyne: A Modern History* (Chichester, 2001); Robert Colls, *Identity of England* (Oxford, 2002); David Matless, *Landscape and Englishness* (London, 1998).
4 John Gloag and Jack Pritchard, 'A Plan for the DIA', *DIA Quarterly Journal*, 17 (January 1932), pp. 7–9.
5 *Enterprise Scotland, 1947* (Edinburgh, 1947), p. 1.
6 Pevsner, *The Englishness of English Art*, p. 186.
7 'Editorial Notes', *DIA Quarterly Journal*, 8 (July 1929), pp. 5 and 6.
8 *Sunday Times Colour Section* (January 1962), p. 1.
9 Gilbert, Matless and Short, *Geographies of British Modernity*, p. 10.
10 A. Greenwood, 'Legislation and the DIA', *DIA Quarterly Journal*, 10 (December 1929), p. 4.
11 'Editorial Notes', *DIA Quarterly Journal*, 9 (October 1929), p. 7.
12 Russell, *Looking North*, p. 5.
13 Ian Cox, *Festival of Britain, South Bank Exhibition: A Guide to the Story It Tells* (London, 1951), p. 9.
14 Ibid., p. 68.
15 Advertisement, ibid., p. v.

1 Modernity and Tradition: Late Victorian and Edwardian Design

1 Jane Beckett and Deborah Cherry, eds, *The Edwardian Era* (London, 1987), p. 8.
2 L. Walker, 'Architecture and Design: Heart of Empire/Glorious Garden, Class and Gender in Edwardian Britain', in *The Edwardian Era*, ed. Beckett and Cherry, p. 118.
3 A. E. Coombes, 'The Franco-British Exhibition: Packaging Empire in Edwardian England', in *The Edwardian Era*, ed. Beckett and Cherry, p. 152.
4 Ibid.
5 D. Feldman, 'Nationality and Ethnicity', in *Twentieth-Century Britain: Economic, Social and Cultural Change*, ed. P. Johnson (London, 1994), pp. 127–48.
6 Ibid., p. 127.
7 B. Schwarz, 'The State', in *The Edwardian Era*, ed. Beckett and Cherry, p. 21.
8 José Harris, *Private Lives, Public Spirit: Britain, 1870–1914* (London, 1993), p. 17.
9 M. Kirby, 'Britain in the World Economy', in *Twentieth-Century Britain*, ed. Johnson, p. 21.
10 Ibid., pp. 33–5.
11 Ibid., p. 36.

12 P. Wardley, 'Edwardian Britain: Empire, Income and Political Discontent', in *Twentieth-century Britain*, ed. Johnson, p. 58.

13 Harris, *Private Lives, Public Spirit*, p. 6.

14 P. Thane, 'The Social, Economic and Political Status of Women', in *Twentieth-Century Britain*, ed. Johnson, p. 95.

15 Ibid., p. 108.

16 Thorstein Veblen, *The Theory of the Leisure Class: An Economic Study of Institutions* (London, 1925), p. 53.

17 See Roszika Parker, 'The Word for Embroidery was WORK', *Spare Rib*, 37 (1975).

18 Christina Hardyment, *From Mangle to Microwave: The Mechanisation of Household Goods* (London, 1988), p. 35.

19 Adrian Forty, *Objects of Desire: Design and Society, 1750–1980* (London, 1986), p. 114.

20 Patricia Branca, *Silent Sisterhood: Middle-class Women in the Victorian Home* (London, 1975), p. 53.

21 Ibid., p. 53.

22 Victoria de Grazia with Ellen Furlough, *The Sex of Things* (Berkeley, CA, 1996), p. 245.

23 Ibid., Table 2, p. 249.

24 Ibid.

25 Ibid., p. 245.

26 Penny Sparke, *As Long as It's Pink* (London, 1995), p. 77.

27 Forty, *Objects of Desire*, p. 115.

28 Hardyment, *From Mangle to Microwave*, p. 187.

29 Sparke, *As Long as It's Pink*, p. 78.

30 Hardyment, *From Mangle to Microwave*, p. 187.

31 Joanna Bourke, *Working-class Cultures in Britain, 1890–1960* (London, 1994), p. 5.

32 Seebohm Rowntree, *Poverty: A Study of Town Life* (London, 1901), p. viii.

33 Ibid., pp. 134–5.

34 Ibid., pp. 269 and 282.

35 Maud Pember Reeves, *Round About a Pound a Week* [1913] (London, 1994).

36 Bourke, *Working-class Cultures in Britain*, pp. 68–9.

37 Ibid., p. 69.

38 Branca, *Silent Sisterhood*, p. 150.

39 David Jeremiah, *Architecture and Design for the Family in Britain, 1900–70* (Manchester, 2000), p. 21.

40 Joseph Rowntree Foundation Archive, York.

41 Peter Davey, *Arts and Crafts Architecture: The Search for Earthly Paradise* (London, 1980), p. 171.

42 Ibid., p. 171.

43 Barry Parker and Raymond Unwin, *The Art of Building a Home* (London, 1901).

44 Matrix, *Making Space: Women and the Man Made Environment* (London, 1984), pp. 69–71.

45 Rowntree, *Poverty*, p. 77.

46 Donald Read, *Documents from Edwardian England* (London, 1973), p. 28.

47 Bayley, *The Garden City* (Milton Keynes, 1975), p. 26.

48 Donald Read, *The Age of Urban Democracy: England, 1868–1914* (London, 1994), p. 401.

49 Bayley, *The Garden City*, p. 35.

50 Davey, *Arts and Crafts Architecture*, p. 9.

51 Frederick Lessore, 'The Art of Reginald Wells, Sculptor and Potter', *Artwork*, II/8 (December–February 1926–7), p. 234.

52 Judy Attfield and Pat Kirkham, *A View From the Interior: Feminism, Women and Design* (London, 1989), p. 172.

53 'Review Notice', *The Studio*, XXVI (1902), p. 101.

54 Elizabeth Cummings, *Phoebe Traquair, 1852–1938* (Edinburgh, 1993), p. 12.

55 Ibid., p. 17.

56 'Glasgow International Exhibition', *The Studio*, XXIII (1901), pp. 239–43.

57 Gavin Stamp, ed., *London 1900* (London, 1978), p. 305.

58 Ibid.

59 For further details, see ibid.

60 Lynne Walker, 'Vistas of Pleasure: Women Consumers of Public Space in the West End of London, 1850–1900', in *Women in the Victorian Art World*, ed. Caroline C. Orr (Manchester, 1995), p. 79.

61 Erika Rappaport, *Shopping for Pleasure: Women in the Making of London's West End* (Princeton, NJ, 2000), p. 149.

62 Ailsa Golding, 'Going Up: Visual Culture and the Department Store in Late Nineteenth-Century Britain (with Particular Reference to the Photographs of Henry Bedford Lemere)', MA dissertation, Northumbria University, 2004, p. 52. See also Kathryn A. Morrison, *English Shops and Shopping: An Architectural History* (New Haven, CT, and London, 2003).

63 Benwell Community Project Final Report Series 6, *The Making of a Ruling Class: Two Centuries of Capital Development on Tyneside* (Newcastle upon Tyne, 1978), p. 42.

64 Ibid., p. 23.

65 Ibid., p. 6.

66 Thomas Faulkner and Andrew Greg, *John Dobson: Architect of the North East* (Newcastle upon Tyne, 2001), pp. 46–7.

67 Ibid., p. 47.

68 Lorna Poole, 'A Fair Deal', *The Gazette* [John Lewis Magazine] (24 September 1988), p. 783.

69 Ibid., p. 784.

70 Bainbridge & Co. Ltd, 'The Emporium of the North' (booklet), John Lewis Archive, 524/D/1.

71 Bainbridge & Co. Ltd, calendar, 1914, John Lewis Archive, 524/D/1.

72 M. Nava, 'Modernity's Disavowal: Women, the City and the Department Store', in *Modern Times: Reflections on English Modernities*, ed. Mica Nava and Anthony O'Shea (London, 1996), p. 53.

73 'New Premises Opened Today', *Newcastle Journal* (10

October 1913), pp. 84–5.

74 Fenwick Archive, box 23, sale catalogues, SC/38–44.

75 *Northern Echo* (12 July 1905), p. 10.

76 Fenwick Archive, box 28, fashion catalogues F/1–34.

77 Ibid.

78 For a fuller discussion, see Cheryl Buckley and Hilary Fawcett, *Fashioning the Feminine: Representation and Women's Fashion from the Fin de Siècle to the Present* (London, 2002), p. 46.

79 For further discussion, see Cheryl Buckley, *Potters and Paintresses: Women Designers in the Pottery Industry, 1870–1955* (London, 1990).

80 Michael Moss and Alison Turton, *A Legend of Retailing: House of Fraser* (London, 1989), p. 67.

81 Coombes, 'The Franco-British Exhibition', in *The Edwardian Era*, ed. Beckett and Cherry, p. 152.

82 Ibid.

83 *Festival of Empire Imperial Exhibition Pageant of London, 1911*, under patronage of His Majesty's Government, Crystal Palace, 1911, p. 7. For a detailed discussion, see Deborah S. Ryan, 'Staging the Imperial City: The Pageant of London, 1911', in *Imperial Cities: Landscape, Display and Identity*, ed. Felix Driver and David Gilbert (Manchester, 1999).

84 *Festival of Empire Imperial Exhibition Pageant of London, 1911*, p. 7.

85 Undated newspaper cutting from Noel, M.P. Scrapbook Containing Materials relating to the Pageant of London, 1911, in National Art Library, Victoria and Albert Museum, London.

86 Harris, *Private Lives, Public Spirit*, p. 255.

87 *Festival of Empire Imperial Exhibition Pageant of London, 1911*, p. 22.

88 Richard Pankhurst, *Sylvia Pankhurst: Artist and Crusader* (London, 1979), pp. 100–1.

89 Ibid., p. 101.

90 For further discussion of these themes, see Lisa Tickner, *The Spectacle of Women: Imagery of the Suffrage Campaign, 1907–14* (London, 1987).

91 Ibid., p. 123.

92 Undated newspaper cutting from Noel, M. P. Scrapbook Containing Materials relating to the Pageant of London, 1911, in National Art Library, Victoria and Albert Museum.

2 'Englishness' and Identity: Design in Early Twentieth-century Britain

1 A.J.P. Taylor, *English History, 1914–1945* (Oxford, 1965), pp. 67–8.

2 Cathy Ross, *Twenties London: A City in the Jazz Age* (London, 2003), pp. 67–8.

3 Ibid., p. 83.

4 Ibid., p. 28.

5 Pamphlet, *The Pottery Trade*, two addresses given at Stoke-on-Trent, January 1918, by H. T. Smith, p. 5.

6 Jon Lawrence, 'The First World War and its Aftermath', in *Twentieth-century Britain: Economic, Social and Cultural Change*, ed. P. Johnson (London, 1994), p. 167.

7 See, for example, Charles Loch Mowat, *Britain Between the Wars, 1918–1940* (Cambridge, 1987).

8 Dudley Baines, 'The Onset of Depression', in *Twentieth-century Britain*, ed. Johnson, p. 169.

9 John Stevenson, *British Society, 1914–45* (London, 1984), p. 103.

10 Baines, 'The Onset of Depression', p. 179.

11 Taylor, *English History*, p. 226.

12 Buckley and Fawcett, *Fashioning the Feminine*, chapter 4.

13 James B. Jefferys, *Retail Trading in Britain, 1850–1950* (Cambridge, 1954), p. 305.

14 Ibid., p. 307.

15 Quoted in Cheryl Buckley and Lynne Walker, *Between the Wars: Architecture and Design on Tyneside, 1919–1939*, exh. cat., Newcastle Polytechnic Art Gallery (Newcastle upon Tyne, 1982), p. 4.

16 John Benson, *The Rise of Consumer Society in Britain, 1880–1980* (London, 1994), p. 144.

17 Ibid., p. 154.

18 *DIA Quarterly Journal*, 9 (October 1929), p. 3.

19 Promotional brochure for Reyrolle Electrics, Hebburn, Tyneside, 1927.

20 *Britannia and Eve* (December 1926), back cover.

21 *Britannia and Eve* (June 1929), p. 131.

22 Ibid., p. 130.

23 Bowes Museum Archive, Barnard Castle, County Durham.

24 Ibid.

25 Editorial, 'Oriental Art', *Burlington Magazine*, XVII (April–September 1910), p. 3.

26 For example, there was an exhibition on blue and white Chinese porcelain at the Burlington Fine Arts Club in 1895.

27 Murray wrote: 'The point of the pot is that the tiger is not of the "Burning Bright" breed, but a benign cat-like pet, pushing its purring way round the pot, as a cat might against one's legs. I remember brushing the tiger on, and thinking of the great cat, and the pot full of milk and its milk-coloured glaze.' See Malcolm Haslam, *William Staite Murray* (London, 1984), p. 73.

28 Basil Gray, 'The Development of Taste in Chinese Art in the West 1872 to 1972', *Transactions of the Oriental Ceramic Society* (1971–2, 1972–3), pp. 21–2.

29 R. L. Hobson, 'Early Staffordshire Wares Illustrated in the British Museum, I', *Burlington Magazine*, II (June–August 1903), p. 64.

30 Ibid., p. 69.

31   Hobson, 'Early Staffordshire Wares Illustrated in the British Museum, v', p. 383.

32   S. K. Tillyard, *The Impact of Modernism, 1900–1920: Early Modernism and the Arts and Crafts* (London, 1988), p. 51.

33   See L. Solon, 'The Lowestoft Porcelain Factory and the Chinese Porcelain Made for the European Market during the Eighteenth Century', *Burlington Magazine*, II (June–August 1903), p. 271.

34   R. L. Hobson, 'Wares of the Sung and Yuan Dynasties, I', *Burlington Magazine*, XV (April–September 1909), p. 23.

35   Roger Fry, *The Burlington Magazine Monograph on Chinese Art* (London, 1925).

36   Tillyard, *The Impact of Modernism*, p. 127.

37   Christopher Green, ed., *Art Made Modern: Roger Fry's Vision of Art*, exh. cat., Courtauld Gallery, London (2000), p. 184.

38   Roger Fry, 'The Art of Pottery in England', *Burlington Magazine*, XXIV (October 1913–March 1914), p. 330.

39   Ibid., p. 335.

40   Ibid.

41   Tillyard, *The Impact of Modernism*, p. 64.

42   Writing to Duncan Grant in 1913 just before Omega opened, Fry asks him to come back soon because there was much to do, including 'china to be painted'. See Denis Sutton, *Letters to Roger Fry, Volume Two* (London, 1972), p. 371.

43   For further information, see Muriel Rose, *Artist Potters in England* (London, 1955), p. 24.

44   Cheryl Buckley, *Potters and Paintresses: Women Designers in the Pottery Industry, 1870–1955* (London, 1990), p. 85.

45   Tanya Harrod, *The Crafts in Britain in the 20th Century* (New Haven, CT, 1999), p. 9.

46   C. Stephens, 'Ben Nicholson: Modernism, Craft and the English Vernacular', in *The Geographies of Englishness: Landscape and the National Past, 1880–1940*, ed. David Peters Corbett, Ysanne Holt and Fiona Russell (New Haven, CT, 2002), pp. 244–5.

47   Haslam, *William Staite Murray*.

48   William Staite Murray, 'Pottery from the Artist's Point of View', *Artwork*, 1/4 (May–August 1925), p. 202.

49   Ibid.

50   Haslam, *William Staite Murray*, p. 27.

51   P. W. Gay and R. I. Smyth, *The British Pottery Industry* (London, 1974), pp. 9–10.

52   Alfred Powell, 'New Wedgwood Pottery', *The Studio*, XCVIII (1929), p. 879.

53   *Arts Etruriae Renascuntur: A Record of the Historical Old Pottery Works of Messrs. Josiah Wedgwood & Sons Ltd. Etruria, England*, 1920.

54   Ibid., p. 3.

55   Ibid., p. 35.

56   Stefan Muthesius, 'Why Do We Buy Old Furniture? Aspects of the Authentic Antique in Britain, 1870–1910',

Art History, XI (June 1988), pp. 238–41.

57   For example, Percy Macquoid, *History of English Furniture: Age of Oak, Age of Walnut, Age of Mahogany, Age of Satinwood*, 4 vols (London, 1904–8).

58   *Country Life*, 904 (2 May 1914), pp. 617–47.

59   *Country Life*, 938 (26 December 1914), pp. 853–6.

60   John Cornforth, *The Search for a Style: Country Life and Architecture, 1897–1935* (London, 1988), p. 12.

61   Ibid., p. 13; *Nostell Priory*, The National Trust (London, 1982).

62   Cornforth, *The Search for a Style*, p. 27.

63   Ibid., p. 27.

64   'Country Homes: Gardens Old and New: Stowe–I', *Country Life*, 887 (3 January 1914), pp. 18–26; 'Country Homes: Gardens Old and New: The Drum', *Country Life*, 979 (9 October 1915), p. 491; and 'Country Homes: Castle-town', *Country Life*, 1131 (7 September 1918), p. xxvii.

65   Percy Macquoid, 'English Tables from 1600–1800, I', *Country Life*, 1154 (15 February 1919), pp. 174–7; 'English Tables from 1600–1800, II', *Country Life*, 1156 (1 March 1919), pp. 228–31; and 'English Tables from 1600–1800, III', *Country Life*, 1158 (15 March 1919), pp. 286–90.

66   'Cookery by Electricity–II', *Country Life*, 1227 (10 July 1920), pp. 69–70; 'Dormanstown', *Country Life*, 1228 (17 July 1920), pp. 97–8.

67   'Lesser Country Houses of Today: Hownhall', *Country Life*, 1242 (23 October 1920), pp. 549–50.

68   Company pamphlet, *Decorative Contracts* by Waring and Gillow, *c.* 1907, National Art Library, Victoria and Albert Museum, London, p. 183.

69   Ibid., p. 191.

70   Company pamphlet, *Adaptations of Antique Furnishings Suitable for Modern Requirements* by Warings, *c.* 1910, The British Library.

71   Company pamphlet, *Past and Present*, Waring and Gillow Ltd, *c.* 1924, The National Art Library, Victoria and Albert Museum, London.

72   Ibid., p. 1.

73   Ibid., pp. 3–5.

74   Ibid., p. 30.

75   Company pamphlet, *Modern Art in French and English Decoration and Furniture*, Waring and Gillow, 1928, The National Art Library, Victoria and Albert Museum, London, p. 2.

76   Ibid.

77   Betty J. Blum, 'Serge Chermayeff', Art Institute of Chicago website (2001), pdf, p. 102: http://www.artic.edu/aic/libraries/caohp/chermayeff.html (accessed 17 January 2006).

78   Company pamphlet, *Suggestions for Spring Furnishings*, Waring and Gillow, 1927, National Art Library, Victoria and Albert Museum, London, p. 43.

79   Company pamphlet, *Illustrated Catalogue of Furniture*

Carpets Bedding and Every Requisite for the Home,
Whiteley's, undated, National Art Library, Victoria and
Albert Museum, London, inner front cover.

80 DIA Quarterly Journal, 4 (July 1928), p. 4.

81 Design and Industry pamphlet, 'A Proposal for the
Foundation of a Design and Industries Association',
undated, British Library, London, p. 3.

82 Design and Industry pamphlet, 'Design in British
Industry', undated, British Library, London, p. 26.

83 Ibid.

84 Sarah Riddick, Pioneer Studio Pottery: The Milner White
Collection (London, 1990), p. 59.

85 Moira Vincentelli, Women and Ceramics: Gendered Vessels
(Manchester, 2000), p. 144.

86 Harrod, The Crafts in Britain, p. 44.

87 Vincentelli, Women and Ceramics, p. 146.

88 Stephen Calloway, Twentieth-century Decoration: The
Domestic Interior from 1900 to the Present Day (London,
1988), p. 143.

89 Design and Industry pamphlet, 'The Pottery Trade',
January 1918, British Library, London, p. 4.

90 'Great Stores and Great Artists', DIA Quarterly Journal, 8
(July 1929), p. 6.

91 A. P. Simon, 'Textiles and Modern Dress', DIA Quarterly
Journal, 4 (July 1928), p. 5.

92 'Advertising the Nation's Workshop', DIA Quarterly
Journal, 8 (July 1929), p. 9.

93 Christopher Reed, Bloomsbury Rooms: Modernism,
Subculture and Domesticity (New Haven, CT, 2004), p. 2.

3 'Going Modern, but Staying British': Design and
Modernisms, 1930–1950

1 Paraphrase of the title of Paul Nash, 'Going Modern and
Being British', published in Weekend Review (12 March
1932) and quoted by J. Collins, 'The Englishness of
English Art', in Modern Britain, 1929–1939, ed. James Peto
and Donna Loveday, exh. cat., Design Museum, London
(1999), p. 69.

2 David Matless, 'Ages of English Design: Preservation,
Modernism and Tales of their History, 1926–1939',
Journal of Design History, III/4 (1990), pp. 203–12.

3 Alison Light, Forever England: Femininity, Literature and
Conservatism between the Wars (London, 1991), p. 10.

4 Unit 1, exh. cat., Portsmouth City Museum and Art
Gallery (Portsmouth, 1978), p. 1.

5 James Peto and Donna Loveday, eds, Modern Britain,
1929–1939, exh. cat., Design Museum, London (1999) p. 11.

6 A seminal example of this is Nikolaus Pevsner's Pioneers
of Modern Design (London, 1936).

7 See 'Art Schools and Industry II: Stoke-on-Trent', The
Studio,103 (1932), pp. 275–83.

8 'Art and Technical Progress in the Potteries', Pottery
Gazette and Glass Trade Review (2 June 1930), pp. 974–6.

9 See Cheryl Buckley, Potters and Paintresses: Women
Designers in the Pottery Industry, 1870–1955 (London, 1990).

10 Margot Coatts, A Weaver's Life, 1872–1952 (Bath, 1983).

11 Tanya Harrod, The Crafts in Britain in the 20th Century
(New Haven, CT, 1999), p. 112.

12 David Peters Corbett, Ysanne Holt and Fiona Russell,
The Geographies of Englishness: Landscape and the National
Past, 1880–1940 (New Haven, CT, 2002), p. 228.

13 Charles Loch Mowat, Britain between the Wars, 1918–1940
(Cambridge, 1987), p. 480.

14 J. B. Priestley, English Journey (London, 1949), p. 397.

15 Ibid., p. 398.

16 Ibid., p. 401.

17 Ibid., pp. 397–8.

18 Ibid., p. 399.

19 Ibid., p. 400.

20 Ibid., p. 402.

21 Ibid., p. 403.

22 Ibid., p. 401.

23 Ibid., p. 411.

24 S. Bowden, 'The New Consumerism', in Twentieth-century
Britain: Economic, Social and Cultural Change, ed. Paul
Johnson (London, 1994), p. 247.

25 Ibid., p. 248. Average consumption of electricity per
domestic consumer was 861 kilowatt hours in the south-
east and 386.4 kilowatt hours in the north-east.

26 Ibid., p. 255.

27 Caroline Haslett, ed., The Electrical Handbook for Women
(London, 1936), p. 15.

28 Ibid., p. 282.

29 Joanna Bourke, Working-class Cultures in Britain,
1890–1960 (London, 1994), p. 15.

30 Elizabeth Roberts, Women and Families, 1940–1970
(Oxford, 1995), p. 23.

31 Priestley, English Journey, p. 314.

32 P. Howlett, 'The War Economy', in Twentieth-century
Britain, ed. Johnson, p. 283.

33 See Judy Attfield, Utility Reassessed: The Role of Ethics in
the Practice of Design (Manchester, 1999).

34 Betty J. Blum, 'Serge Chermayeff', Art Institute of
Chicago website (2001), pdf,
http://www.artic.edu/aic/libraries/caohp/chermayeff.ht
ml (accessed 17 January 2006).

35 Susan Lambert, Paul Nash as Designer, exh. cat., Victoria
and Albert Museum, London (1975); A. Causey, 'Paul
Nash as Designer' in Modern Britain, 1929–1939, ed. Peto
and Loveday, pp. 107–16.

36 Unit 1, p. 20.

37 Lambert, Paul Nash as Designer, p. 12.

38 Causey, 'Paul Nash as Designer', p. 116.

39 Wells Coates, 'Response to Tradition', Architectural

Review (November 1932), p. 168.

40 Ibid., p. 167.

41 Ibid.

42 Ibid.

43 Ibid., p. 168.

44 Ibid., p. 167.

45 Laura Cohn, *Wells Coates: Architect and Designer, 1895–1958*, exh. cat., Oxford Polytechnic (Oxford 1979), p. 5. See also Alastair Grieve, *Isokon* (London, 2004).

46 Wells Coates, 'Furniture Today – Furniture Tomorrow', *Architectural Review* (July 1932), p. 34.

47 Ibid., p. 32.

48 'BBC Studios, Newcastle', *Architects' Journal* (28 February 1935), pp. 329–31; 'Working Details, BBC Studios, Newcastle, Wells Coates', *Architects' Journal* (7 March 1935), pp. 373–6.

49 Cohn, *Wells Coates*, p. 30.

50 Particularly significant writers on this issue were Gordon Forsyth and Nikolaus Pevsner. Although Pevsner's *An Enquiry into Industrial Arts in England* (Cambridge, 1937) included an important discussion of the shortcomings of decoration in the ceramic industry, it was mainly Forsyth who developed the arguments of modernism in relation to the pottery industry in his book *20th Century Ceramics* (London, 1936). Forsyth was widely respected as a designer and educator based in Stoke-on-Trent.

51 'Buyer's Notes', *Pottery Gazette and Glass Trade Review* (1 October 1932), p. 1251.

52 Ibid., p. 1249.

53 For a discussion of these ideas, see, for example, Mark Wigley's essay in D. Fausch, *Architecture, in Fashion* (New York, 1994).

54 'Buyer's Notes', *Pottery Gazette and Glass Trade Review* (1 August 1935), pp. 703–6.

55 *Miners' Welfare Fund Annual Reports 1921–26*, 1 (1926), p. 6.

56 Letter and notes dated 4 October 1984 from R. Bronwyn Thomas ARIBA to Cheryl Buckley.

57 Ibid.

58 Goronwy Rees, *St Michael: A History of Marks and Spencer* (London, 1969), pp. 74–5. See also Rachel Worth, *Fashion for the People: A History of Marks and Spencer* (Oxford, 2006).

59 Goronwy Rees, *St Michael*, p. 119.

60 See Sally Wood, *A Sort of Dignified Flippancy: Penguin Books, 1935–1960* (Edinburgh, 1985); Evelyne Green, *Penguin Books: The Pictorial Cover, 1960–1980* (Manchester, 1981); Steven Hare, *Penguin Portrait: Allen Lane and the Penguin Editors, 1935–1970* (London, 1995); Ruari McLean, *Jan Tschichold: A Life in Typography* (New York, 1997); Phil Baines, *Penguin by Design, 1935–2005* (London, 2005).

61 Robin Kinross, 'Emigré Graphic Designers in Britain: Around the Second World War and Afterwards', *Journal of Design History*, III/1 (1990), p. 52.

62 Gertrude Williams, *The New Democracy: Women and Work* (London, 1945), p. 10.

63 Mowat, *Britain between the Wars*, p. 452

64 Bowden, 'The New Consumerism', p. 256.

65 Lord Gorell, *Art and Industry: Report of the Committee Appointed by the Board of Trade under the Chairmanship of Lord Gorell on the Production and Exhibition of Articles of Good Design and Everyday Use* (London, 1932), p. 14.

66 Ibid., p. 14.

67 Ibid., p. 18.

68 *British Art in Industry, 1935: Illustrated Souvenir*, exh. cat., Royal Academy, London (1935), p. 1.

69 William Rothenstein, 'Possibilities for the Improvement of Industrial Art in England', *Journal of the Royal Society of Arts*, 69 (1921), pp. 268–77; Harold W. Sanderson, 'Art Schools and Art in Industry', *Journal of the Royal Society of Arts*, 82 (1933), pp. 183–99.

70 *British Art in Industry, 1935*, p. 1.

71 Ibid.

72 Ibid.

73 Ibid.

74 John de la Valette, *The Conquest of Ugliness* (London, 1935), advertised in *British Art in Industry, 1935*, p. xx. For further discussion of such literature, see Grace Lees Maffei, 'From Service to Self-Service: Advice Literature as Design Discourse, 1920–1970', *Journal of Design History*, XIV/3 (2001), pp. 187–206, and Jonathan Woodham, 'Managing British Design Reform, II: The Film Deadly Nightshade: As Ill-Fated Episode in the Politics of "Good Taste"', *Journal of Design History*, IX/2 (1996), pp. 101–15.

75 Margaret Bulley, *Have You Good Taste? A Guide to the Appreciation of the Lesser Arts* (London, 1933).

76 Gorell, *Art and Industry*.

77 Ibid., p. 44.

78 Ibid., p. 45.

79 Ibid.

80 Ibid.

81 Duncan Miller, *Interior Decorating: 'How To Do It' Series, Number 13* (London, 1937); Bulley, *Have You Good Taste?*

82 Ibid., p. 1.

83 Ibid., p. 11.

84 Ibid., p. 1.

85 Ibid., p. 19.

86 Miller, *Interior Decorating*, p. 7.

87 Ibid., p. 8.

88 Report by the Council for Art and Industry, *The Working Class Home: Its Furnishing and Equipment* (London, 1937).

89 Ibid., p. 7.

90 Ibid., pp. 9–20.

91 Ibid., p. 46.

92 Ibid., p. 42.

93 Ibid.
94 M. Denney, 'Utility Furniture and the Myth of Utility, 1943–48', in *Utility Reassessed: The Role of Ethics in the Practice of Design*, ed. Judy Attfield (Manchester, 1999), p. 120. See also *cc41 Utility Furniture and Fashion*, exh. cat., Geffrye Museum, London (1974).
95 *cc41 Utility Furniture and Fashion*, p. 15.
96 H. Reynolds, 'The Utility Garment: Its Design and Effect on the Mass Market', in *Utility Reassessed*, ed. Attfield, p. 125.
97 P. Kirkham, 'Fashion, Femininity and "Frivolous" Consumption in World-War-Two Britain', in *Utility Reassessed*, ed. Attfield, p. 147.
98 A. Partington, 'The Designer Housewife in the 1950s', in *A View from the Interior: Feminism, Women and Design*, ed. Judy Attfield and Pat Kirkham (London, 1989), p. 207.
99 *Hansard*, cccvii (26 February 1948), quoted in Helen Crane, 'The Critical Reception of the New Look', MA thesis, Northumbria University (2004), p. 26.
100 Ibid., p. 30.
101 *The Council of Industrial Design*, first annual report, 1945–6 (London), p. 6.
102 Ibid.
103 Ibid.
104 Ibid.
105 Draft report, 'The Training of the Industrial Designer for Industry', 26 February 1946, p. 6.
106 Ibid., p. 5.
107 'Design Centre for the Potteries?', *Staffordshire Sentinel* (10 December 1947).
108 See Penny Sparke, ed., *Did Britain Make It? British Design in Context, 1946–86* (London, 1986); Patrick J. Maguire and Jonathan M. Woodham, *Design and Cultural Politics in Postwar Britain: The Britain Can Make It Exhibition of 1946* (Leicester, 1997); Becky E. Conekin, '*The Autobiography of a Nation': The 1951 Festival of Britain* (Manchester, 2003).
109 Maguire and Woodham, *Design and Cultural Politics in Postwar Britain*, p. 57.
110 *A Pictorial Review of Scottish Industry as Displayed in the Exhibition Enterprise Scotland*, exh. cat., Council of Industrial Design (London, 1947), p. 1.
111 Maguire and Woodham, *Design and Cultural Politics in Postwar Britain*, p. 60.
112 *The Council of Industrial Design*, third annual report, 1947–8 (London), p. 3.
113 Ibid., p. 3.
114 Gordon Russell and Jacques Groag, *The Story of Furniture* (West Drayton, 1947), p. 2.
115 Ibid., p. 2.
116 Ibid.
117 Ibid., p. 23.
118 Ibid., p. 30

119 John Gloag, *The English Tradition in Design* (London, 1947), p. 36.
120 Ibid., p. 35.
121 Nikolaus Pevsner, *Pioneers of Modern Design* (London, 1936).
122 Gloag, *The English Tradition*, p. 15.
123 Ibid., pp. 16–17
124 Ibid., p. 36.

4 Designing the 'Detergent Age': Design in the 1950s and '60s

1 Ian Cox, *Festival of Britain, South Bank Exhibition: A Guide to the Story It Tells* (London, 1951), p. 9.
2 C. R. Schenk, 'Austerity and Boom', in *Twentieth-century Britain: Economic, Social and Cultural Change*, ed. P. Johnson (London, 1994), p. 309.
3 Reyner Banham, *Theory and Design in the First Machine Age* (London, 1960), p. 9.
4 Ibid., p. 10.
5 Ibid., p. 9.
6 Millicent Frances Pleydell-Bouverie, *Daily Mail Book of Britain's Post-War Homes* (London, 1944), p. 12.
7 Chris Stephens and Katharine Stout, eds, *Art and the 60s: This Was Tomorrow* (London, 2004), p. 15.
8 Ibid., p. 10.
9 Ibid.
10 Tanya Harrod, *The Crafts in Britain in the 20th Century* (New Haven, CT, 1999), p. 232.
11 Ibid., p. 244.
12 B. Curtis, 'A Highly Mobile and Plastic Environ', in *Art and the 60s*, ed. Stephens and Stout, p. 48.
13 Angela McRobbie, *British Fashion Design: Rag Trade or Image Industry?* (London, 1998), p. 35.
14 Liz Heron, ed., *Truth, Dare or Promise: Girls Growing Up in the 50s* (London, 1985), p. 2.
15 Ibid., p. 156.
16 Ibid.
17 Elizabeth Roberts, *Women and Families: An Oral History, 1940–1970* (Oxford, 1995), p. 29.
18 Alan Sked and Chris Cook, *Post-War Britain: A Political History. New Edition, 1945–1992* (London, 1993), p. 123.
19 P. J. Madgwick, D. Seeds and L. J. Williams, *Britain since 1945* (London, 1982), p. 13.
20 Ibid., p. 13.
21 Schenk, 'Austerity and Boom', p. 318.
22 Madgwick, Seeds and Williams, *Britain since 1945*, p. 35.
23 T. Kushner, 'Immigration and "Race Relations" in Postwar British Society', in *Twentieth-century Britain*, ed. Johnson, p. 413.
24 Paul Gilroy, *There Ain't No Black in the Union Jack* (London, 1987), p. 79.

25 G. Lewis, 'From Deepest Kilburn', in *Truth, Dare or Promise*, ed. Heron, p. 223.

26 Carol Tulloch, 'There's No Place Like Home: Home Dressmaking and Creativity in the Jamaican Community of the 1940s to the 1960s', in *The Culture of Sewing: Gender, Consumption and Home Dressmaking*, ed. Barbara Burman (Oxford, 1999), p. 122.

27 Gertrude Williams, *Women and Work: The New Democracy* (London, 1945), p. 9.

28 Ibid.

29 Ibid.

30 Women between the ages of 20 and 64. Data from Pat Thane, 'Women since 1945', in *Twentieth-century Britain*, ed. Johnson, p. 393.

31 Jane Lewis, *Women in Britain since 1945* (Oxford, 1992), p. 67.

32 Ibid., p. 68.

33 Thane, 'Women since 1945', p. 400.

34 Lewis, *Women in Britain since 1945*, p. 43.

35 Ibid., p. 44.

36 Mary Banham and Bevis Hillier, *A Tonic to the Nation: The Festival of Britain, 1951* (London, 1976), p. 190.

37 Ibid., p. 191.

38 Becky E. Conekin, *'The Autobiography of a Nation': The 1951 Festival of Britain* (Manchester, 2003), p. 33.

39 Ibid., p. 9. See also Anne Massey, *The Independent Group: Modernism and Mass Culture in Britain, 1945–1959* (Manchester, 1995), p. 10.

40 Conekin, *'The Autobiography of a Nation'*, p. 10.

41 Ibid., p. 131.

42 Banham and Hillier, *A Tonic to the Nation*, p. 40.

43 Ibid., p. 114.

44 Ibid., p. 26.

45 Ibid., p. 194.

46 Ibid., p. 197.

47 Cox, *Festival of Britain*, p. 74.

48 Lionel Brett, *The Things We See: Houses* (Middlesex, 1947), p. 6.

49 Banham and Hillier, *A Tonic to the Nation*, p. 193.

50 Ibid.

51 Lesley Jackson, *The New Look: Design in the Fifties* (London, 1991).

52 For a detailed discussion of the debates about the Royal College of Art during this period, see Christopher Frayling, *The Royal College of Art: One Hundred and Fifty Years of Art and Design* (London, 1987).

53 See chapter seven in Harrod, *The Crafts in Britain*.

54 Ibid., p. 232.

55 Ibid.

56 Edward de Waal, *20th century Ceramics* (London, 2003), p. 127.

57 Harrod, *The Crafts in Britain*, p. 250.

58 Andrew Greg, *Primavera: Pioneering Craft and Design,*

1945–1995, exh. cat., Tyne and Wear Museums, Newcastle (1995), p. 4.

59 Pleydell-Bouverie, *Daily Mail Book of Britain's Post-War Homes*, p. 23. For detailed discussion of the Ideal Homes exhibitions, see Deborah S. Ryan, *The Ideal Home through the 20thc.* (London, 1997).

60 Pleydell-Bouverie, *Britain's Post-War Homes*, pp. 13–14.

61 Ibid., p. 22.

62 Ibid., plan, insert before p. 25.

63 Ibid., p. 47.

64 Leaflet, *Northern Housing exhibition*, Baths Hall, Northumberland Road, Newcastle upon Tyne (1–30 December 1944), p. 8.

65 Leaflet, *The Gas Industry Presents . . . 'The Practical Northern Home'*, 1944.

66 Ibid.; for a fuller discussion of this, see also Lynne Walker, *Women Architects: Their Work* (London, 1984).

67 Pleydell-Bouverie, *Britain's Post-War Homes*, p. 130.

68 Banham and Hillier, *A Tonic to the Nation*, p. 60.

69 Nigel Whiteley, *Pop Design: Modernism to Mod* (London, 1987), p. 41.

70 Jonathan Woodham, *Twentieth-century Design* (Oxford, 1997), p. 190.

71 *Ward's Directory, Newcastle upon Tyne*, 1938.

72 Author interview with Ian Caller, 21 December 2006.

73 Plans T186/15642 Tyne and Wear Museums Archive.

74 Author interview with Ian Caller, 21 December 2006.

75 Ibid.

76 Ibid.

77 Ibid.

78 See also 'Why Not a Chair of Paper or Foam?', *Newcastle Evening Chronicle* (21 March 1969), p. 9.

79 'Design Centre Comes North', *Newcastle Journal* (20 March 1969), p. 6.

80 Author interview with Ian Caller, 21 December 2006.

81 Ibid.

82 Ibid.

83 Massey, *The Independent Group*, p. 98.

84 Simon Sadler, 'British Architecture in the Sixties', in *Art and the 60s*, ed. Stephens and Stout, p. 124, and 'Design at the Design Museum: Alison and Peter Smithson', http://www.designmuseum.org/design/index (accessed 10 September 2005).

85 Sadler, 'British Architecture in the Sixties', p. 131, and 'Design at the Design Museum: Archigram', http://www.designmuseum.org/design/index (accessed 10 September 2005).

86 For further discussion, see P. Usherwood ' Art on the Margins', in *Newcastle upon Tyne: A Modern History*, ed. Robert Colls and Bill Lancaster (Chichester, Sussex, 2001), pp. 245–66.

87 David Robbins, ed., *The Independent Group: Postwar Britain and the Aesthetics of Plenty* (Cambridge, MA, 1990),

p. 133.

88 Peter Coe, *Lubetkin and Tecton: Architecture and Social Commitment* (London, 1981), p. 38.

89 Ibid., p. 186.

90 Ibid.

91 Graeme Rigby and Sally-Ann Norman, 'Farewell Squalor', website of Side Photographic Gallery, Side Collection, The Side, Newcastle upon Tyne, http://www.amber–online.com/gallery/exhibition215/notes215.html (accessed 15 October 2005).

92 Ibid.

93 Ibid.

94 The Twentieth Century Society, Casework Reports: Pasmore Pavilion, Sunny Blunts, Peterlee, Co. Durham (Victor Pasmore, 1967–70), website http://www.c20society.org.uk/docs/casework/pasmore.html (accessed 15 October 2005).

95 Thomas E. Faulkner, 'Conservation and Renewal in Newcastle upon Tyne', in *Northumbrian Panorama: Studies in the History and Culture of North East England* (London, 1996), p. 140.

96 For a fuller discussion, see T. E. Faulkner 'Architecture in Newcastle' and D. Byrne, 'The Reconstruction of Newcastle: Planning since 1941', in *Newcastle upon Tyne*, ed. Colls and Lancaster.

97 Faulkner, 'Architecture in Newcastle', p. 239.

98 Byrne, 'The Reconstruction of Newcastle', p. 347.

99 N. Vall, 'The Emergence of the Post-Industrial Economy in Newcastle, 1914–2000', in *Newcastle upon Tyne*, ed. Colls and Lancaster, p. 66.

100 Paddy Maguire, 'Craft Capitalism and the Projection of British Industry in the 1950s and 1960s', *Journal of Design History*, VI/2 (1993), p. 101.

101 Dick Hebdige, *Hiding in the Light* (London, 1998), p. 65.

102 Dick Hebdige, *Sub-culture: The Meaning of Style* (London, 1979), p. 29.

103 C. Tulloch, 'Rebel Without a Pause: Black Street style and Black Designers', in *Chic Thrills a Fashion Reader*, ed. Juliet Ash and Elizabeth Wilson (London, 1992), p. 86. See also Carol Tulloch, ed., *Black Style* (London, 2004).

104 Hebdige, *Sub-culture*, p. 40.

105 Tulloch 'Rebel Without a Pause', p. 89.

106 Jonathon Green, *All Dressed Up: The Sixties and the Counter Culture* (London, 1999), p. x.

107 Ibid., p. 3.

108 Hebdige, *Sub-culture*, p. 50; Christopher Breward, *Fashioning London: Clothing and the Modern Metropolis* (Oxford, 2004), p. 133. See also Christopher Breward, David Gilbert and Jenny Lister, *Swinging Sixties: Fashion in London and Beyond, 1955–1970* (London, 2006).

109 Nik Cohn quoted in Green, *All Dressed Up*, p. 9.

110 C. Evans 'Post-War Poses, 1955–75', in *The London Look: Fashion from Street to Catwalk*, ed. Christopher Breward,

Edwina Ehrman and Caroline Evans (New Haven, CT, 2004), p. 118.

111 Breward, *Fashioning London*, p. 155.

112 For further discussion, ibid., p. 152, McRobbie, *British Fashion Design*, p. 37.

113 Barbara Bernard, *Fashion in the 60s* (London, 1978), pp. 15–16.

114 Green, *All Dressed Up*, p. 76.

115 H. Radnor, 'On the Move: Fashion Photography and the Single Girl in the 1960s', in *Fashion Cultures: Theories, Explorations and Analysis*, ed. Stella Bruzzi and Pamela Church Gibson (London, 2000). See also Judith Watt, *Ossie Clark, 1965–74* (London, 2003).

116 F. Jameson, 'Postmodernism and Consumer Society', in *Postmodern Culture*, ed. Hal Foster (London, 1983), p. 111.

5 The Ambiguities of Progress: Design from the Late 1960s to 1980

1 Raphael Samuel, *Theatres of Memory* (London, 1994), p. 83.

2 'All You Need Is Love: The Beatles in Pepperland', *Sunday Times Magazine* (14 July 1968), pp. 40–41; 'Nothing in Common but Paris', *Sunday Times Magazine* (25 August 1968), p. 23; 'Ginger Rogers: Still Dreaming the American Dream', *Sunday Times Magazine* (22 December 1968), pp. 18–34; 'Design For Living', *Sunday Times Magazine* (1 December 1968), pp. 18–21.

3 V. Papanek, *Design for the Real World: Human Ecology and Social Change* (London, 1974).

4 'About Us' page, 'Foundation Principles', Traidcraft website: http://www.traidcraft.co.uk/template2.asp?pageID=1643&fromID=1275 (accessed 1 November 2005).

5 Nigel Whiteley, *Design for Society* (London, 1993), p. 17.

6 Peter York, *Style Wars* (London, 1980), p. 12.

7 David Harvey, *The Condition of Postmodernity* (Oxford, 1989), p. 9.

8 Ibid., p. 9.

9 Elizabeth Roberts, *Women and Families: An Oral History, 1940–1970* (Oxford, 1995), p. 21.

10 Ibid.

11 Jane Lewis, *Women since 1945* (Oxford, 1992), p. 2.

12 Roberts, *Women and Families*, p. 21.

13 For examples, see Lewis, *Women since 1945*, and P. J. Madgwick, D. Steeds and L. J. Williams, *Britain since 1945* (London, 1982).

14 Arthur Marwick, *British Society since 1945* (London, 1982), p. 188.

15 Ibid., p. 203.

16 Ibid.

17 Ibid., p. 189.

18 Madgwick, Steeds and Williams, *Britain since 1945*, p. 345.

19 T. Kushner, 'Immigration and "Race Relations" in Postwar British Society', in *Twentieth-Century Britain:*

Economic, Social and Cultural Change, ed. P. Johnson (London, 1994), p. 419.

20 Ibid.

21 Paul Gilroy, *There Ain't No Black in the Union Jack: The Cultural Politics of Race and Nation* (London, 1987), p. 119.

22 Ibid., p. 120.

23 Ibid., p. 122.

24 Charles Jencks, *The Language of Post-modern Architecture* (London, 1991), p. 24.

25 Ibid., p. 13.

26 Lyall in *The State of British Architecture*, p. 43, illustrates the demolition of Oak and Eldon Gardens in Birkenhead in 1979.

27 Special issue, 'The North', *Sunday Times Magazine* (28 March 1965), pp. 15–41.

28 Janet Stewart, 'The Drift to the North', in ibid., p. 33.

29 David Beal, 'The Great Face-Lift', in ibid., p. 40.

30 Stewart, 'The Drift to the North', p. 32.

31 Ibid., p. 33.

32 Thomas E. Faulkner, *Northumbrian Panorama: Studies in the History and Culture of North East England* (London, 1996), p. 144.

33 Lyall, *The State of British Architecture*, p. 137.

34 'The Byker Way Forward: Byker – Issues Arising from the "Listing" Proposal', *A Report for Newcastle City Council* (January 2001), p. 4.

35 Jencks, *The Language of Post-modern Architecture*, p. 13.

36 Nigel Whiteley, *Pop Design: Modernism to Mod* (London, 1987), p. 88.

37 Penny Sparke, *An Introduction to Design and Culture in the Twentieth Century* (London, 1986), p. 143.

38 'All You Need Is Love', pp. 40–41; 'All Wrapped Up for Summer', *Sunday Times Magazine* (15 June 1975).

39 Ernestine Carter, 'Into the Big Time', *Sunday Times Colour Section* (4 February 1962), p. 20.

40 'Design for Living: A Revival for the Gothick Revival', *Sunday Times Colour Section* (29 July 1962), pp. 15–19.

41 'Design for Living: The Sunday Times Family Kitchen', *Sunday Times Colour Section* (15 July 1962), p. 17.

42 Matrix, *Making Space: Women and the Man Made Environment* (London, 1984), p. 106.

43 Stephen Bayley, *Taste: The Secret Meaning of Things* (London, 1991), p. 192.

44 Ann Oakley in *Housewife* (London, 1976).

45 Christopher Frayling, *The Royal College of Art: One Hundred and Fifty Years of Art and Design* (London, 1987), p. 154.

46 Priscilla Chapman, 'Design For Living: What's the Matter with British Furniture', *Sunday Times Magazine* (31 January 1965), p. 43.

47 Barty Phillips, *Conran and the Habitat Story* (London, 1984), p. 27.

48 'Cartel Storage Units', Habitat Advertisement in *Sunday Times Magazine* (26 September 1971), p. 10.

49 Whiteley, *Pop Design*, p. 123.

50 Terence Conran, *The House Book* (London, 1974), p. 8.

51 Ibid., p. 9.

52 Ibid.

53 See Naomi Games, Catherine Moriarty and June Rose, *Abram Games Graphic Designer: Maximum Meaning, Minimum Means* (London, 2003); Jeremy Aynsley, *A Century of Graphic Design* (London, 2001); Philip B. Meggs, *A History of Graphic Design* (Chichester, 1998); Wally Olins, *The New Guide to Identity: How to Create and Sustain Change through Managing Identity. Wolff Olins* (London, 1995); Wally Olins, *Corporate Identity: Making Business Strategy Visible through Design* (London, 1989).

54 Peter Dormer, *Design since 1945* (London, 1993), p. 112.

55 'Hello. I'm Wally', website: http://wallyolins.com/ (accessed 17 June 2006). See Rick Poyner, ed., *Independent British Graphic Design since the Sixties* (London, 2004).

56 Arguing that the counter-culture was both cultural and political, Jonathon Green implied that the former was the more effective: *All Dressed Up: The Sixties and the Counterculture* (London, 1999), p. 113.

57 Tanya Harrod, *The Crafts in Britain in the 20th Century* (New Haven, CT, 1999), p. 241.

58 Green, *All Dressed Up*, pp. 164–5.

59 Frayling, *The Royal College of Art*, p. 185.

60 Green, *All Dressed Up*, p. xi.

61 B. Curtis, 'A Highly Mobile and Plastic Environ', in *Art and the 60s: This Was Tomorrow*, ed. Chris Stephens and Katharine Stout (London, 2004), p. 58.

62 Harrod, *The Crafts in Britain*, p. 241.

63 Ibid.

64 Barbara Hulanicki, *From A to Biba* (London, 1984), p. 98.

65 Vance Packard, *The Hidden Persuaders* (New York, 1957), and Ralph Nader, *Unsafe at Any Speed: The Designed-In Dangers of the American Automobile* (New York, 1966).

66 Jonathan Woodham, *Twentieth-Century Design* (Oxford, 1997), p. 230.

67 'About Us' page, History of Oxfam, Oxfam website: http://www.oxfam.org.uk/about_us/history/index.htm (accessed 12 November 2005).

68 'Ergonomics in Human Sciences – An Introduction' page, 'The *Torrey Canyon* Supertanker Disaster', Department of Human Sciences, Loughborough University, website: http://www.lboro.ac.uk/departments/hu/ergsinhu/aboutergs/torrey.html (accessed 11 November 2005).

69 Carl Gardner and Julie Sheppard, *Consuming Passion: The Rise of Retail Culture* (London, 1989), p. 222.

70 'Our Values' page, The Body Shop, website: http://www.thebodyshopinternational.com/web/tbsgl/values.jsp (accessed 14 November 2005).

71 'BBC Body Shop Opinion', 'Have Your Say: Is Body Shop Takeover Worth it?', published 17 March 2006, http://newsdorums.bbc.co.uk/nol/thread (accessed 16 June 2006).

72 Penny Sparke, *Italian Design, 1870 to the Present* (London, 1988).

73 Woodham, *Twentieth-Century Design*, p. 191.

74 Sparke, *Italian Design, 1870 to the Present*, p. 162.

75 Hilary Wainwright and Dave Elliott, *The Lucas Plan* (London, 1982).

76 For further information, see Pauline Madge, 'Design, Ecology, Technology: A Historiographical Review', *Journal of Design History*, VI/ 3 (1993), pp. 149–66.

77 Papanek, *Design in the Real World*, p. ix.

78 Ibid., p. xviii.

79 'About Us' page, 'Foundation Principles', Traidcraft website: http://www.traidcraft.co.uk/template2.asp?pageID=1643&fromID=1275 (accessed 14 November 2005).

80 Ibid.

81 Christopher Breward, *Fashioning London: Clothing and the Modern Metropolis* (Oxford, 2004), p. 184.

82 Samuel, *Theatre of Memory*, p. 67.

83 'Cecil Beaton's Dream Wardrobe', *Sunday Times Magazine* (5 September 1971), pp. 14–19.

84 M. Battersby, *The Decorative Twenties* (London, 1969); *The Decorative Thirties* (London, 1971).

85 Mo Teitelbaum, 'Lady of the Rue Bonaparte', *Sunday Times Magazine* (22 June 1975), pp. 28–40; for another example on the work of inter-war designer Ashley Havinden, see William Feaver, 'Walk the Ashley Way', *Sunday Times Magazine* (6 April 1975), pp. 72–5.

86 An approach exemplified in Sheila Rowbotham, *Hidden from History* (London, 1973).

87 Frederic Jameson, 'Postmodernism and Consumer Society', in *Postmodern Culture*, ed., H. Foster (London, 1983), p. 125.

88 Ibid., pp. 124–5.

89 Ibid., p. 114.

90 Angela McRobbie, ed., *Zoot Suits and Second-hand Dresses: An Anthology of Fashion and Music* (London, 1989), p. 48. For further discussion, see Nicky Gregson and Louise Crewe, *Second-hand Cultures* (Oxford, 2003).

91 Dick Hebdige, 'A Report on the Western Front: Postmodernism and the "Politics" of Style', in *The BLOCK Reader in Visual Culture*, ed. Jon Bird et al. (London, 1996), p. 297.

92 Ibid., pp. 297–8.

93 It began in 1964 in Abingdon Road, moved in 1966 to Kensington Church Street, then to Kensington High Street in 1969, with one further move on the same street to a former department store, Derry and Toms, in 1973.

94 E. Wilson, 'Biba's Style of Femininity', in BIBA: *The Label, The Lifestyle, The Look*, exh. cat., Tyne and Wear Museums, Newcastle upon Tyne (1993), p. 8.

95 A. Massey, 'BIBA: Interior Lifestyles', in ibid., p. 23

96 C. McDermott, 'The Biba Logo: The First Designer Lifestyle', in ibid., p. 18.

97 C. Ross, 'Biba, Black Dwarf, Black Magic Women', in ibid., p. 14.

98 Wilson, 'Biba's Style of Femininity', pp. 9–10.

99 Breward, *Fashioning London*, p. 174.

100 Hulanicki, *From A to Biba*, p. 98.

101 Ibid.

102 Ibid., p. 63.

103 Ibid., p. 98.

104 Ross, 'Biba, Black Dwarf, ' in BIBA, exh. cat., p. 13.

105 Hulanicki, *From A to Biba*, p. 100.

106 McRobbie, ed., *Zoot Suits*, p. 34.

107 C. Evans, 'Cultural Capital, 1976–2000', in *The London Look: Fashion from Street to Catwalk*, ed. Christopher Breward, Edwina Ehrman and Caroline Evans (New Haven, CT, 2004), p. 141.

108 D. Hebdige, *Sub-culture: The Meaning of Style* (London, 1979), p. 67.

109 Ibid., p. 64.

110 Carol Tulloch, 'Rebel Without a Pause: Black Street Style and Black Designers', in *Chic Thrills: A Fashion Reader*, ed. Juliet Ash and Elizabeth Wilson (London, 1992), p. 90.

111 Ibid.

112 Caroline Evans and Minna Thornton, *Women and Fashion: A New Look* (London, 1989), p. 18.

113 Ibid., p. 24.

114 Hebdige, *Sub-culture*, p. 107.

115 Teal Trigs, 'Scissors and Glue: Punk Fanzines and the Creation of a DIY Aesthetic', *Journal of Design History*, XIX/1 (2006), pp. 69–83.

6 'I Shop Therefore I Am': Design since the '80s

1 Title of an artwork by Barbara Kruger, 1983.

2 R. Hewison, 'Commerce and Culture', in *Enterprise and Heritage: Crosscurrents of National Culture*, ed. John Corner and Sylvia Harvey (London, 1991), p. 163.

3 Stephen Bayley, *Commerce and Culture: From Pre-Industrial Art to Post-Industrial Value* (London, 1989).

4 Ibid., p. 5.

5 Hewison, 'Commerce and Culture', p. 173.

6 John Corner and Sylvia Harvey, eds., *Enterprise and Heritage: Crosscurrents of National Culture* (London, 1991), p. 9.

7 Assertion by the British Prime Minister, Margaret Thatcher, in an interview in *Woman's Own* (23 September 1987).

8 Corner and Harvey, eds, *Enterprise and Heritage*, p. 15.

9 Robert Hewison, *The Heritage Industry: Britain in a*

Climate of Decline (London, 1987).

10 Ibid., p. 9.

11 Alan Sked and Chris Cook, *Post-war Britain: A Political History. New Edition, 1945–1992* (London, 1993), p. 516.

12 L. Hannah, 'Crisis and Turnaround? 1973–1993', in in *Twentieth-century Britain: Economic, Social and Cultural Change*, ed. P. Johnson (London, 1994), p. 348.

13 Sked and Cook, *Post-war Britain*, p. 584. There were consecutive Conservative Governments in the years 1979–83, 1983–7, 1987–92, 1992–7.

14 Ibid.

15 Ibid., pp. 584–5.

16 Corner and Harvey, eds, *Enterprise and Heritage*, p. 11.

17 Ibid., p. 14.

18 K. Robins, 'Tradition and Translation: National Culture in its Global Context', in *Enterprise and Heritage*, p. 34.

19 'About Beamish: What Is Beamish?' page, Beamish website: http://www.beamish.org.uk/about.html (accessed 22 January 2006).

20 Ibid.

21 'Archive of Labour Manifestos' website: 1997 Manifesto pages, http://www.labour–party.org.uk/manifestos/1997/1997–labour–manifesto.shtml (accessed 24 January 2006).

22 'New Left Review Homepage', article by S. Watkins, 'A Weightless Hegemony: New Labour's Role in the Neoliberal Order', *New Left Review*, no. 25 (January–February 2005), http://www.newleftreview.net/NLR25901.shtml (accessed 24 January 2006).

23 Ibid.

24 Robert Lumley, ed., *The Museum Time Machine* (London, 1988), p. 1.

25 Ibid.

26 Ibid., p. 2.

27 'About the Building and its Architect' page, Imperial War Museum North website: http://north.iwm.org.uk/server/show/ConWebDoc.993 (accessed 28 November 2005).

28 Ibid.

29 F. Bianchini and H. Schwengel, 'Re-imagining the City', in *Enterprise and Heritage*, p. 219.

30 Robins, 'Tradition and Translation', in *Enterprise and Heritage*, p. 34.

31 T. Bennett, 'Museums and "the People"', in *The Museum Time Machine*, p. 73.

32 Robins, 'Tradition and Translation', p. 36.

33 For a discussion of this, see John Thackera, ed., *Design After Modernism* (London, 1988), p. 14.

34 Bianchini and Schwengel, 'Re-imagining the City', p. 219.

35 'LSE Discussion Papers page', London School of Economics website, pdf; Mark Kleinman, '"A More Normal Housing Market?": The Housing Role of the London Docklands Development Corporation,

36 Deyan Sudjic, *Cult Objects* (London, 1985).

37 'About Us' page, The Design Museum website, pdf: 'About the Design Museum', www.designmuseum.org/httpd/flash/about.html (accessed 29 November 2005); and Hilary Cottam, http://www.design–council.org.uk/webdav/servlet/ (accessed 27 June 2006).

38 'Concorde', www.designmuseum.org/design/index (accessed 27 June 2006).

39 See Rick Poyner, ed., *Independent British Graphic Design since the Sixties* (London, 2004). See also 'Jonathan Ive', *Creative Review*, XVII/26 (26 January 1997).

40 British Council, 'Artists' website: http://collection.britishcouncil.org/html/work/work.aspx?a=1&id=45591&section=/artist/ (accessed 27 June 2006).

41 B. West, 'The Making of the English Working Past: A Critical View of the Ironbridge Gorge Museum', in *The Museum Time Machine*, p. 37.

42 Ibid.

43 Jonathan Woodham, *Twentieth-century Design* (Oxford, 1997), p. 215.

44 Frances Hannah, *Ceramics* (London, 1986); Kathy Niblett, *Dynamic Design: The British Pottery Industry, 1940–1990* (Stoke-on-Trent, 1990), and 'A–Z of Stoke-on-Trent Potters', Staffordshire Potteries Ltd' page, the potteries.org website, http://www.thepotteries.org/allpotters/949.htm (accessed 22 January 2006).

45 My thanks to Andrew Casey for providing further information and images on this.

46 Sutherland Lyall, *The State of British Architecture* (London, 1980), p. 74.

47 Ibid., pp. 79–80.

48 'About Us' page and 'Current Developments' page (includes link to Leconfield Sales Brochure pdf), Bussey & Armstrong website: http://www.busseyarmstrong.co.uk/index.htm (accessed 1 December 2005).

49 Ibid.

50 Advertisement for 'Leconfield' in *Darlington Today*, issue 2 (2005), pp. 22–3, and 'About Us' page and 'Current Developments' page (includes link to Leconfield Sales Brochure pdf), Bussey & Armstrong website: http://www.busseyarmstrong.co.uk/index.htm (accessed 1 December 2005).

51 Ibid.

52 Peter York and Charles Jennings, *Peter York's Eighties* (London, 1995), p. 77.

53 Ibid., p. 88.

54 Hewison, *The Heritage Industry*, p. 66.

55 Nikolaus Pevsner, *The Englishness of English Art* (London, 1955), p. 185.

56 Ibid., p. 186.

57 Angela McRobbie, *British Fashion Design: Rag Trade or Image Industry?* (London, 1998), p. 110.

58 R. Arnold, 'Vivienne Westwood's Anglomania', in *The Englishness of English Dress*, ed. Christopher Breward, Becky Conekin and Caroline Cox (Oxford, 2002), p. 167.

59 McRobbie, *British Fashion Design*, p. 110.

60 A. Ribeiro, 'On Englishness in Dress', in *The Englishness of English Dress*, p. 25.

61 Ibid.

62 Catherine McDermott, *Made in Britain: Tradition and Style in Contemporary British Fashion* (London, 2002), p. 6.

63 David Harvey, *The Condition of Postmodernity* (Oxford, 1989), p. 53.

64 Ibid., p. 54.

65 Lynne Walker, ed., *Drawing on Diversity. Women, Architecture and Practice*, exh. cat., RIBA Heinz Gallery, London (1997), p. 25.

66 Matrix, *Making Space: Women and the Man-made Environment* (London, 1984), p. vii.

67 Mark Swennarton, 'Matrix', *Building Design*, 940 (9 June 1989), p. 4.

68 Ibid.

69 Ruth Owens, 'Child Care Challenge', *Architects' Journal* (18 October 1989), pp. 38–45.

70 York and Jennings, *Peter York's Eighties*, p. 56.

71 Carl Gardner and Julie Shepherd, *Consuming Passion: The Rise of Retail Culture* (London, 1989), p. 84.

72 Ibid.

73 McRobbie, *British Fashion Design*, pp. 47–8.

74 *Next Directory*, no. 1 (1988), p. 76.

75 Rowena Chapman and Jonathan Rutherford, eds, *Male Order: Unwrapping Masculinity* (London, 1988), p. 23.

76 Frank Mort, *Cultures of Consumption: Masculinities and Social Space in Late Twentieth Century Britain* (London, 1996), p. 116.

77 Chapman and Rutherford, eds, *Male Order*, p. 59.

78 Ibid. See Rick Poyner, ed., *Independent British Graphic Design since the Sixties* (London, 2004).

79 See Penny Sparke, *An Introduction to Design and Culture: 1900 to the Present* (London, 2004), p. 214.

80 Robins, 'Tradition and Translation', in *Enterprise and Heritage*, p. 27.

81 'Jonathan Ive, Designer of the Year 2003' page, The Design Museum website: http://www.designmuseum.org/ design/index.php?id=63 (accessed 22 January 2006).

82 For further discussion of technology-led design, gender and consumption, see Paul Atkinson, 'Man in a Briefcase: The Social Construction of the Laptop Computer and the Emergence of a Type Form', *Journal of Design History*, XVIII/2 (2005), pp. 191–205.

83 Jon Wozencroft, *The Graphic Language of Neville Brody 2* (London, 1994), p. 110. See Poyner, ed., *Independent British Graphic Design since the Sixties*.

84 *Next Directory*, no. 1 (1988), p. 3.

85 Ibid., p. 308.

86 Penny Sparke, *Italian Design: 1870 to the present* (London, 1988), p. 225.

87 Gardner and Shepherd, *Consuming Passion*, p. 21.

88 IKEA catalogue (2002), p. 9.

89 Elen Lewis, *Great IKEA! A Brand for All the People* (London, 2005), p. 123.

90 See Sparke, *An Introduction to Design and Culture*, p. 206.

91 Bianchini and Schwengel, 'Re-imagining the City', p. 212.

92 Gardner and Shepherd, *Consuming Passion*, p. 183.

93 Figures from N. Vall, 'The Emergence of the Post-industrial Economy in Newcastle, 1914–2000', in *Newcastle upon Tyne: A Modern History*, ed. Robert Colls and Bill Lancaster (Chichester, 2001), p. 61.

94 Gardner and Shepherd, *Consuming Passion*, p. 41.

95 'Metro Centre Student Info' page, pdf Metro Centre Student Info pacl, Metro Centre website: http://www.metrocentre–gateshead.co.uk/ (accessed 6 December 2005).

96 Ibid.

97 Bianchini and Schwengel, 'Re-imagining the City', p. 220.

98 Vall, 'The Emergence of the Post-industrial Economy', p. 70.

Select Bibliography

A305 *History of Architecture and Design, 1890–1939* (Milton Keynes, 1975)

A Pictorial Review of Scottish Industry as displayed in the exhibition, Enterprise Scotland, Council of Industrial Design, exh. cat. (London, 1947)

Anderson, Benedict, *Imagined Communities: Reflections on the Origin and Spread of Nationalism* (London, 1983)

Arts Etruriae Renascuntur: A Record of the Historical Old Pottery Works of Messrs Josiah Wedgwood & Sons Ltd, Etruria, England, 1920

Ash, Juliet, and Elizabeth Wilson, eds, *Chic Thrills: A Fashion Reader* (London, 1992)

Attfield, Judy, *Utility Reassessed: The Role of Ethics in the Practice of Design* (Manchester, 1999)

—, and Pat Kirkham, *A View from the Interior: Feminism, Women and Design* (London, 1989)

Aynsley, Jeremy, *A Century of Graphic Design* (London, 2001)

Baines, Phil, *Penguin By Design, 1935–2005* (London, 2005)

Banham, Mary, and Bevis Hillier, *A Tonic to the Nation: The Festival of Britain 1951* (London, 1976)

Banham, Reyner, *Theory and Design in the First Machine Age* (London, 1960)

Battersby, Martin, *The Decorative Twenties* (London, 1969)

Bayley, Stephen, *The Garden City* (Milton Keynes, 1975)

—, *Commerce and Culture: From Pre-Industrial Art to Post-Industrial Value* (London, 1989)

—, *Taste: The Secret Meaning of Things* (London, 1991)

Beckett, Jane, and Deborah Cherry, eds, *The Edwardian Era* (London, 1987)

Benson, John, *The Rise of Consumer Society in Britain, 1880–1980* (London, 1994)

Benton, Charlotte, Tim Benton and Ghislaine Wood, eds, *Art Deco, 1910–1939*, exh. cat., Victoria and Albert Museum, London (2003), p. 261

Benwell Community Project Final Report Series 6, *The Making of a Ruling Class: Two Centuries of Capital Development on Tyneside* (Newcastle upon Tyne, 1978).

Bernard, Barbara, *Fashion in the 60s* (London, 1978)

BIBA: *The Label, The Lifestyle, The Look*, exh. cat., Tyne and Wear Museums, Newcastle upon Tyne (1993)

Bird, Jon, et al., eds, *The BLOCK Reader in Visual Culture* (London, 1996)

Blaszczyk, Regina L., *Imagining Consumers: Design and Innovation from Wedgwood to Corning* (Baltimore, MD, and London, 2000)

Bourke, Joanna, *Working-class Cultures in Britain, 1890–1960* (London, 1994)

Boyce, David G., *Decolonisation and the British Empire, 1775–1997* (London, 1999)

Branca, Patricia, *Silent Sisterhood: Middle-class Women in the Victorian Home* (London, 1975)

Brett, Lionel, *The Things We See: Houses* (Middlesex, 1947)

Breward, Christopher, *Fashion* (Oxford, 2003)

—, *Fashioning London: Clothing and the Modern Metropolis* (Oxford, 2004)

—, Becky Conekin and Caroline Cox, eds, *The Englishness of English Dress* (Oxford, 2002)

—, Edwina Ehrman and Caroline Evans, eds, *The London Look: Fashion from Street to Catwalk* (New Haven, CT, 2004)

—, David Gilbert and Jenny Lister, *Swinging Sixties: Fashion in London and Beyond, 1955–1970* (London, 2006)

British Art in Industry, 1935: Illustrated Souvenir, exh. cat. Royal Academy, London (1935)

Bruzzi, Stella, and Pamela Church Gibson, eds, *Fashion Cultures: Theories, Explorations and Analysis* (London, 2000)

Buckley, Cheryl, *Potters and Paintresses: Women Designers in the Pottery Industry, 1870–1955* (London, 1990)

—, and Hilary Fawcett, *Fashioning the Feminine: Representation and Women's Fashion from the Fin de Siècle to the Present* (London, 2002)

—, and Lynne Walker, eds, *Between the Wars: Architecture and Design on Tyneside, 1919–1939*, exh. cat., Newcastle

Polytechnic Art Gallery (Newcastle upon Tyne, 1982).

Bulley, Margaret, *Have You Good Taste? A Guide to the Appreciation of the Lesser Arts* (London, 1933)

Burke, Peter, *Varieties of Cultural History* (Oxford, 1997).

Burman, Barbara, ed., *The Culture of Sewing: Gender, Consumption and Home Dressmaking* (Oxford, 1999)

Calloway, Stephen, *Twentieth-century Decoration: The Domestic Interior from 1900 to the Present Day* (London, 1988)

cc41 Utility Furniture and Fashion, exh. cat. (London, 1974)

Chapman, Rowena, and Jonathan Rutherford, eds, *Male Order: Unwrapping Masculinity* (London, 1988)

Clark, Hazel and Alexandra Palmer, eds, *Old Clothes, New Looks: Second-hand Fashion* (Oxford, 2005)

Coatts, Margot, *A Weaver's Life, 1872–1952* (Bath, 1983)

Coe, Peter, *Lubetkin and Tecton: Architecture and Social Commitment* (London, 1981)

Cohn, Laura, *Wells Coates: Architect and Designer, 1895–1958*, exh. cat., Oxford Polytechnic (Oxford, 1979)

Colls, Robert, *Identity of England* (Oxford, 2002)

—, and Philip Dodd, eds, *Englishness: Politics and Culture, 1880–1920* (London, 1986)

—, and Bill Lancaster, eds, *Newcastle upon Tyne: A Modern History* (Chichester, 2001)

Conekin, Becky E., *'The Autobiography of a Nation': The 1951 Festival of Britain* (Manchester, 2003)

Corbett, David P., Ysanne Holt and Fiona Russell, eds, *The Geographies of Englishness: Landscape and the National Past, 1880–1940* (New Haven, CT, 2002)

Corner, John, and Sylvia Harvey, eds, *Enterprise and Heritage: Crosscurrents of National Culture* (London, 1991)

Cornforth, John, *The Search for a Style: Country Life and Architecture, 1897–1935* (London, 1988)

Cox, Ian, ed., *Festival of Britain, South Bank Exhibition: A Guide to the Story It Tells* (London, 1951)

Cummings, Elizabeth, *Phoebe Traquair, 1852–1938* (Edinburgh, 1993)

Davey, Peter, *Arts and Crafts Architecture: The Search for Earthly Paradise* (London, 1980)

Dormer, Peter, *Design since 1945* (London, 1993)

Douglas, M., and B. Isherwood, *The World of Goods: Towards an Anthropology of Consumption* (London, 1979)

Eagleton, Terry, *The Idea of Culture* (Oxford, 2000)

Evans, Caroline, and Minna Thornton, *Women and Fashion: A New Look* (London, 1989)

Faulkner, Thomas, ed., *Northumbrian Panorama: Studies in the History and Culture of North East England* (London, 1996)

—, and Andrew Greg, *John Dobson: Architect of the North East* (Newcastle upon Tyne, 2001)

Faulkner, Thomas, Peter Babcock and Paul Jones, *Newcastle and Gateshead: Architecture and Heritage* (Liverpool, 2006)

Fausch, Debra, *Architecture in Fashion* (New York, 1994)

Forsyth, Gordon, *20th Century Ceramics* (London, 1936)

Forty, Adrian, *Objects of Desire: Design and Society, 1750–1980* (London, 1986)

Foster, H., ed., *Postmodern Culture* (London, 1983)

Frayling, Christopher, *The Royal College of Art: One Hundred and Fifty Years of Art and Design* (London, 1987)

Games, Naomi, Catherine Moriarty and June Rose, *Abram Games, Graphic Designer: Maximum Meaning, Minimum Means* (London, 2003)

Gardner, Carl, and Julie Sheppard, *Consuming Passion: The Rise of Retail Culture* (London, 1989)

Gay, P. W., and R. I. Smyth, *The British Pottery Industry* (London, 1974)

Gilbert, David, David Matless and Brian Short, *Geographies of British Modernity: Space and Society in the Twentieth Century* (Oxford, 2003)

Gilroy, Paul, *There Ain't No Black in the Union Jack* (London, 1987)

Gloag, John, *The English Tradition in Design* (London, 1947)

Gorell, Lord, *Art and Industry: Report of the Committee Appointed by the Board of Trade under the Chairmanship of Lord Gorell on the Production and Exhibition of Articles of Good Design and Every-day Use* (London, 1932).

de Grazia, Victoria, with Ellen Furlough, eds, *The Sex of Things* (Berkeley, CA, 1996)

Green, Christopher, ed., *Art Made Modern: Roger Fry's Vision of Art*, exh. cat., Courtauld Gallery, London (2000)

Green, Jonathon, *All Dressed Up: The Sixties and the Counter Culture* (London, 1999)

Greg, Andrew, *Primavera: Pioneering Craft and Design, 1945–1995*, exh. cat., Tyne and Wear Museums (Newcastle upon Tyne, 1995)

Gregson, Nicky and Louise Crewe, *Second-hand Cultures* (Oxford, 2003)

Hannah, Frances, *Ceramics* (London, 1986)

Hardyment, Christina, *From Mangle to Microwave: The Mechanisation of Household Goods* (London, 1988)

Harris, José, *Private Lives, Public Spirit: Britain, 1870–1914* (London, 1993)

Harrod, Tanya, *The Crafts in Britain in the 20th Century* (New Haven, CT, 1999)

Harvey, David, *The Condition of Postmodernity* (Oxford, 1989)

Haslam, Malcolm, *William Staite Murray* (London, 1984)

Haslett, Caroline, ed., *The Electrical Handbook for Women* (London, 1936)

Hebdige, Dick, *Sub-culture: The Meaning of Style* (London, 1979)

Heron, Liz, *Truth, Dare or Promise: Girls Growing up in the 50s* (London, 1985)

Hewison, Robert, *The Heritage Industry: Britain in a Climate of Decline* (London, 1987)

Hobson, Robert L., *The George Eumorfopoulos Collection: Catalogue of the Chinese, Corean and Persian Pottery and Porcelain, Volume One: From the Chou to the end of the T'ang Dynasty* (London, 1925)

—, Bernard Rackham and W. King, *Chinese Ceramics in Private Collections* (London, 1931)

Holtby, Winifred, *Women and a Changing Civilisation* (London, 1934)

Hulanicki, Barbara, *From A to Biba* (London, 1984)

Huneault, Kristina, *Difficult Subjects: Working Women and Visual*

Culture, Britain, 1880–1914 (Aldershot, 2002)

Jackson, Lesley, *The New Look: Design in the Fifties* (London, 1991)

Jefferys, James B., *Retail Trading in Britain, 1850–1950* (Cambridge, 1954)

Jencks, Charles, *The Language of Post-modern Architecture* (London, 1991)

Jephcott, Pearl A., *Girls Growing Up* (London, 1942)

Jeremiah, David, *Architecture and Design for the Family in Britain, 1900–70* (Manchester, 2000)

Johnson, Paul, ed., *Twentieth-century Britain: Economic, Social and Cultural Change* (London, 1994)

Lambert, Susan, *Paul Nash as Designer*, exh. cat. Victoria and Albert Museum, London (1975)

Legeza, Ireneus László, ed., *A Descriptive and Illustrated Catalogue of the Malcolm Macdonald Collection of Chinese Ceramics in the Gulbenkian Museum of Oriental Art and Archaeology* (Oxford, 1972)

Lewis, Elen, *Great IKEA! A Brand for All the People* (London, 2005)

Lewis, Jane, *Women in Britain since 1945* (Oxford, 1992)

Light, Alison, *Forever England: Femininity, Literature and Conservatism between the Wars* (London, 1991)

Lumley, Robert, ed., *The Museum Time Machine* (London 1988)

Lunn, Richard, *Pottery: A Handbook of Practical Pottery for Art Teachers and Students* (London, 1903)

Lyall, Sutherland, *The State of British Architecture* (London, 1980)

McDermott, Catherine, *Made in Britain: Tradition and Style in Contemporary British Fashion* (London, 2002)

McRobbie, Angela, *British Fashion Design: Rag Trade or Image Industry?* (London, 1998)

—, ed., *Zoot Suits and Second-hand Dresses: An Anthology of Fashion and Music* (London, 1989)

Madgwick, P. J., D. Seeds and L. J. Williams, *Britain since 1945* (London, 1982)

Maguire, Patrick J., and Jonathan M. Woodham, *Design and Cultural Politics in Postwar Britain: The Britain Can Make It Exhibition of 1946* (Leicester, 1997)

Marwick, Arthur, *British Society since 1945* (London, 1982)

Massey, Anne, *The Independent Group: Modernism and Mass Culture in Britain, 1945–1959* (Manchester, 1995)

Matless, David, *Landscape and Englishness* (London, 1998)

Matrix, *Making Space: Women and the Man Made Environment* (London, 1984)

Meggs, Philip B., *A History of Graphic Design* (Chicester, 1998)

Miller, Daniel, *Material Culture and Mass Consumption* (London, 1987)

—, *Interior Decorating: 'How To Do It' Series, Number 13*, Studio (London, 1937)

Morrison, Kathryn A., *English Shops and Shopping: An Architectural History* (New Haven, CT, and London, 2003)

Mort, Frank, *Cultures of Consumption: Masculinities and Social Space In Late Twentieth Century Britain* (London, 1996)

Moss, Michael, and Alison Turton, *A Legend of Retailing: House of Fraser* (London, 1989)

Mowat, Charles Loch, *Britain Between the Wars, 1918–1940* (Cambridge, 1987)

Nader, Ralph, *Unsafe at Any Speed: The Designed-In Dangers of the American Automobile* (New York, 1966)

Nairn, Tom, *The Break-up of Britain: Crisis and Neo-Nationalism* (London, 1977)

Nava, Mica, and Anthony O'Shea, eds, *Modern Times: Reflections on a Century of English Modernity* (London, 1996)

Niblett, Kathy, *Dynamic Design: The British Pottery Industry, 1940–1990* (Stoke-on-Trent, 1990).

Oakley, Ann, *Housewife* (London, 1976)

Orr, Clarissa Campbell, ed., *Women in the Victorian Art World* (Manchester, 1995)

Orwell, George, *Coming Up for Air* (London, 1939)

Packard, Vance, *The Hidden Persuaders* (New York, 1957)

Pankhurst, Richard, *Sylvia Pankhurst: Artist and Crusader* (London, 1979)

Papanek, Victor, *Design for the Real World: Human Ecology and Social Change* (London, 1974)

Pember Reeve Maud, *Round about a Pound a Week* [1913] (London, 1994)

Peto, James, and Donna Loveday, eds, *Modern Britain, 1929–1939*, exh. cat., Design Museum, London (1999)

Pevsner, Nikolaus, *An Enquiry into Industrial Arts in England* (Cambridge, 1937)

—, *The Englishness of English Art* (London, 1955)

—, *Pioneers of Modern Design* (London, 1936)

Phillips, Barty, *Conran and the Habitat Story* (London, 1984)

Phizacklea, Annie, *Unpacking the Fashion Industry,* (London, 1990)

Pleydell-Bouverie, Millicent Frances, *Daily Mail Book of Britain's Post-War Homes* (London, 1944)

Poyner, Rick, ed., *Independent British Graphic Design since the Sixties* (London, 2004)

Priestley, J. B., *English Journey* (London, 1949)

Rappaport, Erika, *Shopping for Pleasure: Women in the Making of London's West End* (Princeton, NJ, 2000)

Read, Donald, *Documents from Edwardian England* (London, 1973)

—, *The Age of Urban Democracy: England, 1868–1914* (London, 1994)

Reed, Christopher, *Bloomsbury Rooms: Modernism, Subculture and Domesticity* (New Haven, CT, 2004)

Rees, Goronwy, *St Michael: A History of Marks and Spencer* (London, 1969)

Report by the Council for Art and Industry, *The Working Class Home: Its Furnishing and Equipment* (London, 1937)

Riddick, Sarah, *Pioneer Studio Pottery: The Milner White Collection* (London, 1990)

Robbins, David, ed., *The Independent Group: Postwar Britain and the Aesthetics of Plenty* (Cambridge, MA, 1990)

Roberts, Elizabeth, *Women and Families: An Oral History, 1940–1970* (Oxford, 1995)

Rose, Muriel, *Artist Potters in England* (London, 1955)

Rowbotham, Sheila, *Hidden From History* (London, 1973)

Rowntree, Seebohm B., *Poverty: A Study of Town Life* (London, 1901)

—, *Poverty and Progress: A Second Survey of York* (London, New York and Toronto, 1941)

Russell, Dave, *Looking North: Northern England and the National Imagination* (Manchester, 2004)

Russell, Gordon, and Jacques Groag, *The Story of Furniture* (London, 1947)

Samuel, Raphael, *Theatres of Memory* (London, 1994)

Sked, Alan, and Chris Cook, *Post-War Britain: A Political History. New Edition, 1945–1992* (London, 1993)

Sparke, Penny, *An Introduction to Design and Culture in the Twentieth Century* (London, 1986)

—, *An Introduction to Design and Culture: 1900 to the Present* (London, 2004)

—, *Italian Design: 1870 to the Present* (London, 1988)

—, *As Long as Its Pink* (London, 1995)

—, ed., *Did Britain Make It? British Design in Context, 1946–86* (London, 1986)

Stephens, Chris, and Katharine Stout, eds, *Art and the 60s: This Was Tomorrow* (London, 2004)

Stevenson, John, *British Society, 1914–45* (London, 1984)

Sutton, Denis, *Letters to Roger Fry: Volume Two* (London, 1972)

Taylor, A.J.P., *English History, 1914–1945* (Oxford, 1965)

Taylor, Avram, *Working Class Credit and Community since 1918* (Basingstoke, 2002)

Thackera, John, ed., *Design After Modernism* (London, 1988)

The Council of Industrial Design, first annual report (London, 1945–6)

The Council of Industrial Design, third annual report (London, 1947–8)

Thirties: British Art and Design before the War, exh. cat. Hayward Gallery, London (1979)

Thompson, Edward P., *The Making of the English Working Class* (London, 1963)

Tickner, Lisa, *The Spectacle of Women: Imagery of the Suffrage Campaign, 1907–14* (London, 1987)

Tillyard, Stella K., *The Impact of Modernism, 1900–1920: Early Modernism and the Arts and Crafts* (London, 1988)

Tulloch, Carol, ed., *Black Style* (London, 2004)

Veblen, Thorstein, *The Theory of the Leisure Class: An Economic Study of Institutions* (London, 1925)

Vincentelli, Moira, *Women and Ceramics: Gendered Vessels* (Manchester, 2000)

de Waal, Edward, *20th Century Ceramics* (London, 2003)

Wainwright, Hilary, and Dave Elliott, *The Lucas Plan* (London, 1982)

Walker, Lynne, *Women Architects: Their Work* (London, 1984)

—, ed., *Drawing on Diversity: Women, Architecture and Practice*, exh. cat., RIBA Heinz Gallery, London (1997)

Watt, Judith, *Ossie Clark, 1965–74* (London, 2003)

Wedgwood, exh. cat., Grafton Galleries, London (1936)

Whiteley, Nigel, *Design for Society* (London, 1993)

—, *Pop Design: Modernism to Mod* (London, 1987)

—, *Reyner Banham: Historian of the Immediate Future* (London, 2002)

Williams, Gertrude, *Women and Work: The New Democracy* (London, 1945)

Wolfe, Tom, *From Bauhaus to Our House* (London, 1989)

Woodham, Jonathan M., *Twentieth-Century Design* (Oxford, 1997)

Wozencroft, Jon, *The Graphic Language of Neville Brody 2* (London, 1994)

Worth, Rachel, *Fashion for the Public: A History of Marks & Spencer* (Oxford, 2006)

York, Peter, *Style Wars* (London, 1980)

—, and Charles Jennings, *Peter York's Eighties* (London, 1995)

Acknowledgements

My thanks to all those who helped in the preparation of this book; they are too numerous to list, but without exception I benefited from their enormous generosity. A few people have been both great friends and wonderful colleagues, reading my work over the years and commenting on it, particularly Lynne Walker, Pat Kirkham, Deborah Cherry, Jane Beckett and Ysanne Holt. Special thanks to Tom Faulkner and to Vivian Constantinopoulos. I am grateful to the School of Arts and Social Sciences at Northumbria University for funding towards the illustrations of this book and for the continued support of my research over many years. This book is dedicated to my family – Allan, Kate, Tom and Anna – who gave me the time and space to get on with it.

Photo Acknowledgements

The author and publishers wish to express their thanks to the following sources of illustrative material and/or permission to reproduce it. Every effort has been made to contact the copyright holders. If there are any inadvertent omissions these will be corrected in a future reprint.

Antique Collectors' Club: pp. 98, 139, 215; Jane Atfield: p. 213 (foot); photo Birmingham City Archives: p. 157; Bussey-Armstrong: p. 216; Country Life Picture Library: pp. 71, 72; Crafts Study Centre, Farnham, Surrey (www.csc.ucreative.ac.uk): pp. 46, 61 (top), 79, 87; Darlington Library Centre for Local Studies: p. 58; David Frater: pp. 153, 154, 170, 204; The Design Archives, Brighton (designarchives@brighton.ac.uk): pp. 118, 119, 127, 144, 179; The Design Museum, London: p. 208; photo André Huber: p. 210; reproduced courtesy of the estate of Abram Games: p. 126; Glasgow University Archives: p. 40; Ian Caller: pp. 148, 149; Pauline Caller: p. 150; IKEA Ltd (www.ikea.co.uk): p. 228; photo Images of London: p. 34 (left); reproduced courtesy of the Trustees of the Imperial War Museum: p. 115; John Lewis Partnership Archive Collection: p. 14; photo John McCann/RIBA Library Photographs Collection: p. 151; The Joseph Rowntree Foundation (Joseph Rowntree Archive): pp. 23, 25, 26; Laura Ashley Archives: p. 193; London's Transport Museum, © Transport for London: pp. 34 (right), 54; Lynne Walker: p. 31; Lynne and Steve Walker p. 186 (right); reproduced courtesy of Marks and Spencer: p. 103; Matrix: p. 221; photo Metro Centre: p. 230; National Motor Museum, Beaulieu: pp. 124, 155; photo Ivan Nemec: p. 211 (left); Newcastle City Council Local Studies Collection: pp. 80, 81; NI Syndication: pp. 160, 172, 173; Oxford Designers & Illustrators: p. 213 (top left); reproduced by permission of Penguin Books Ltd: p. 105; University Library, University of East Anglia, Norwich (Pritchard Papers): pp. 8, 87; RCAHMS (Bedford Lemere Collection): p. 35 (foot); RIBA British Architectural Library Photographs Collection: p. 152; The Scotsman Publications Ltd. Licensor (www.scran.ac.uk): p. 176; The Selfridges Archive at The History of Advertising Trust (www.hatads.org.uk): p. 35 (top); SCRAN/Edward Martin. Licensor (www.scran.ac.uk): p. 213 (top right); Sheffield Galleries & Museums Trust: p. 175; Traidcraft: p. 186 (left); Tyne and Wear Museums: p. 190; photos V&A Images/Victoria & Albert Museum, London: pp. 42, 64, 74, 76, 82, 93, 94; Walter Gumiero/Magis: p. 196; the Wedgwood Museum Archives: pp. 62, 99; the Women's Library, London: pp. 44, 194.

Index

Aalto, Alvar 209

abstraction 7, 9, 59, 60, 63–4, 68, 78–9, 86, 101–2, 123, 129, 150–1, 168, 187

Adam, Robert 70

Adams, John 60

Adshead, Ramsey and Abercrombie 70, 71–2, *71*, *72*

advertising 9–13, 16, 19, 52–4, 86, 90, 93–5, 116, 133, 176, 180, 195, 224–6

agit-prop design 179

Air Chair *196*, 212

Alessi 226, *227*

Amies, Hardy 115

anti-design 183–4, 194, 199

Antiuniversity 180–1

Arad, Ron 211, *211*

Archigram 147, 183

Architectural Review 81, 84, 92–3, 96, 127

architecture 167–8, 187, 216

abstraction 101–2, *102*, 150, 151, 168, 187

and art, integration of 129

classicism 33, 36, 67

designer status 211

diversity 9, 128, 169, 188–90, 203

eclecticism 167, 170–1, 183, 188, 207, 217

end of modern (1972) 167

and heritage 202, 203, 205–7, 218, 230–1

historic premises, adaptation of 203, 204–6

House of the Future (1956) 147, *151*

housing, post-war 136–7, 146–7, 150–3, 167–71

and modernism 95–104, *97*, *102*, *103*, 137, 150, 153, 167, 169–71, 187, 198, 205

and modernity 70, 168–9, 187, 203,

205, 206

neo-vernacular 215–17, *216*

New Architecture Movement 222

owner-occupier housing (inter-war) 90–1, 131

postmodern 170–1, 230

reworking 18th century 67–72, *71*, *72*, 78, 83

and shopping culture, late Victorian and Edwardian period 33–8

urban renewal 206–7, 208

women, for and by 141–3, *142*, 187, *221*, 222–3

Archizoom 183

Arena 225, *226*

Arp, Hans 137

Art Deco 75–6, 83, 92, 94, 95, 161, 187, 189, 190, 191

Art Nouveau 121, 161, 178, 189, 191

Arts and Crafts Movement 15, 22–33, 44, 48, 59, 60–8, 77–8, 81, 85, 94–5, 115, 120–1, 161, 187, 217

Ashbee, C.R. 29

Ashley, Laura 192, *193*, 197, 214

Atfield, Jane 212–14, *213*

Auld, Barbara 141–2

Auld, Ian 129

avant-garde design 147–8, 187

Bakelite *82*, 98–9

Baker, Robert 139

Baldwin, Gordon 129, 140

Banham, Reyner 127–8, 134, 136, 137, 148, 187

Barlow, Thomas 117, 120

Barnsley, Grace 29, 68

Barron, Phyllis 48, 87–8, *87*

Bartlett, David 145

Basic Design 129, 140, 149

Bass, Saul 179, 209

Bates, John 158

Battersby, Martin 187

Bauhaus 9, 81, 83, 85, 129, 140

BBC studios 98, 99

Beamish Museum 202, 206

Beardsley, Aubrey 189

Beattie, William *35*, 36

Bell, Vanessa 65, 66, 68

Bennett, Gloria 132

Bertoia, Harry 137, 145, 146

Biba 162, 189, 190–2, *190*

Billington, Dora 60

black communities *see* immigrants

Black, Misha 137

BOAC/BEA 6, 12

Body Shop 182–3

Bogue, Moira 194

Bonsiepe, Gui 184

book design 94, 95, 104

Bourneville 23

boutiques 153, 157, 158, 185, 186, 189

Braden, Norah 78, 89

Brandt, Edgar 76

Brett, Lionel 137

Breuer, Marcel 83, 88, 98, 177

Britain

1930s 88–92

1970s 163–6

1980s 200–3

and England, distinction between 10–11, 40, 41, 45, 75, 88–92, 135, 214, 219–20